PHYSIOLOGY - LABORATORY AND CLINICAL RESEARCH

GAIT BIOMETRICS

BASIC PATTERNS, ROLE OF NEUROLOGICAL DISORDERS AND EFFECTS OF PHYSICAL ACTIVITY

PHYSIOLOGY - LABORATORY AND CLINICAL RESEARCH

Additional books in this series can be found on Nova's website under the Series tab.

Additional e-books in this series can be found on Nova's website under the e-book tab.

PHYSIOLOGY - LABORATORY AND CLINICAL RESEARCH

GAIT BIOMETRICS

BASIC PATTERNS, ROLE OF NEUROLOGICAL DISORDERS AND EFFECTS OF PHYSICAL ACTIVITY

LI LI
AND
MATTHEW HOLMES
EDITORS

NOVA BIOMEDICAL

New York

For permission to use material from this book please contact us:
Telephone 631-231-7269; Fax 631-231-8175
Web Site: http://www.novapublishers.com

Library of Congress Cataloging-in-Publication Data

ISBN: 978-1-62948-381-8

Library of Congress Control Number: 2013951581

Published by Nova Science Publishers, Inc. † New York

Contents

In: Gait Biometrics
Editors: Li Li and Matthew Holmes
ISBN: 978-1-62948-381-8

Chapter I

Introduction to Gait Biometrics

Barry Munkasy*
Department of Health and Kinesiology, Georgia Southern University,
Statesboro, GA, US

Abstract

Despite basic gait normally being mastered early in life understanding gait still challenges experts. Ways of quantifying movement are constantly being proposed, utilized, and critiqued as others are using these methodologies to better understand the challenges produced by injury, disease, and amputation. Before delving into advanced gait topics, this introductory chapter provides a review of the basic concepts of gait biomechanics.

Despite basic gait normally being mastered early in life understanding gait challenges experts on a day to day basis. Ways of quantifying movement are constantly being proposed, utilized, and critiqued as others are using these methodologies to better understand the challenges produced by injury, disease, and amputation. While some are examining methods to quantify asymmetries and variability others are working to understand the effect of concussions, diabetes, low back pain, spinal cord injury, amputation, and Parkinson's disease on gait.

Before embarking on an advanced examination of the biomechanics of gait a review of basic concepts will hopefully provide a foundation for better understanding.

While gait may be characterized by various forms of ambulation it most normally refers to walking. In humans walking is a bipedal task differentiated from running by a double support (stance) phase. That is a period of time in which both feet are in contact with the ground. This is in contrast to running which characteristically has a period of non-support, neither foot is in contact with the ground. Both walking and running have periods of single

* BMunkasy@GeorgiaSouthern.edu.

support which occurs when only one foot is in contact with the ground. The gait cycle of any single limb is a repeated sequence of swing and support phases. If a foot is not in contact with the ground then it is in swing, otherwise it is in stance phase.

Fundamental understanding of gait begins with an examination of the movement of the body's center of mass. The center of mass is the unique point where the body's mass may be considered to be concentrated. Its location is based upon the mass and the distribution of the mass of the body's segments. Tracking the center of mass in three dimensions summarizes the body's movement. Movement of the center of mass may be referred to as sway and is often examined in the sagittal (looking from either left or right side of the body) and frontal (looking from either the front or the back of the body) planes.

A step may be defined as the event of one foot to the future same event of the contralateral foot. For example, heel strike of the right foot to heel strike of the left foot. A stride is properly defined as two steps or the event of one foot to the future same event of the same foot. However, connotatively the terms step and stride are often thought of as synonyms. Some prefer to use the word cycle instead of stride to refer to two steps and not use the term stride at all. Unfortunately, this practice is not universally accepted and much confusion may result as to which an author refers.

Kinematics is the description of motion. Gait may be characterized from a kinematic perspective based on temporal and spatial aspects of movement. Step time, cycle time, single support time, double support time, non-support time and swing time are useful temporal descriptors of gait. Step length, cycle length, and step width provide useful spatial descriptors of gait. The number of steps per second is referred to as step rate (cycle, cycle rate) or cadence. Gait velocity is defined as displacement divided by time and can be calculated as the product of step length and step rate (cycle length * cycle rate). While technically not the same some refer to gait velocity as gait speed.

The base of support encompasses the area of the body's contact with the ground. Both in static and dynamic circumstances it may be relatively small when one foot is in contact with the ground or large with both feet on the ground (step width). During static situations to be balanced the center of mass must lie over the base of support. During dynamic activities the center of mass may be outside the base of support. Stability is a measure of the amount of effort required to keep the center of mass over the base of support. Wider bases of support are more stable then smaller ones. As gait is initiated the center of mass progresses toward and then over the base of support. When terminated the center of mass is initially outside and then stops within the base of support.

Gait may also be studied from a kinetic perspective. Kinetics examines the cause (forces) of movement and includes the use of force plates and electromyography. Force plates provide a summation of the forces applied by the body to the ground. Normally the equal, but opposite force is presented (force applied to the body by the ground) and referred to as the ground reaction force. The resultant ground force may be characterized by a magnitude, direction, and a point of application (center of pressure). These descriptors change continually during the gait cycle. Three useful components of the resultant ground reaction force are the vertical, anterior/posterior (direction of progression), and medial/lateral (side to side) forces.

Electromyography (EMG) is the study of the electrical activity of the muscles that produce movement. As such EMG provides a window into the internal forces and processes that result from the actions of the central and peripheral nervous system. One aspect of coordination may be assessed by examining the onset and offset of muscle activity. Under special circumstances the amount of EMG may be quantified to provide insight as to the amount of force produced during a particular action.

Both ground reaction forces and kinematics may be combined using inverse dynamics to estimate the forces and torques across joints, power produced and transferred, and work done. The force across the joint may provide an estimate of the axial and sheer forces experienced by bone and ligamentous tissue. Torques (moments) is the measure of the tendency of a force to cause rotation and helps estimate muscles contribution to movement. Work is the product of the displacement that results from force and power is work per unit of time or the product of force and velocity. Kinetics and kinematics may also be combined to examine the coupling and decoupling of the center of pressure and center of mass during the initiation and termination of gait.

While study of spatial and temporal minimums, maximums, and averages of various gait variables has provided a foundation for understanding, newer analyses have focused on the variability of these variables during gait. While some would argue that variability is a hallmark of a compromised system, it appears that there is a range of required variability with too little or too much indicating compromise.

In the following chapters methodology for quantifying gait and the effect of disease and injury are presented. Finally a novel use of gait analysis is presented.

From a methodological perspective Exell in his chapter *A review of developments in gait asymmetry quantification* and Rhea and Kiefer"s *Patterned variability in gait behavior: How can it be measured and what does it mean?* provide background and advances in the quantification of gait. They then discuss potential application to researchers and clinicians.

The effect of disease on gait is discussed by Crews and Yalla in *Biomechanical and physical activity considerations for diabetic peripheral neuropathy* and by Hondzinski, Kadivar, and Hegwood in *Gait in Parkinson's disease*. Each begins by presenting the fundamentals of the disease under consideration, discusses the effect of the disease on gait and then concludes with how treatment and physical activity can improve these conditions.

The effect of relatively permanent changes to the body on gait are presented by Perez-Nombela and colleagues and Vanicek. Perez-Nombela and colleagues in *Gait analysis of two incomplete spinal cord syndromes: Central cord and Brown-Sequard* define these syndromes and describe their effect on gait. They conclude by discussing treatments to improve gait. In *The effects of altered lower limb mechanics as a result of lower limb amputation on gait and posture and the implications for falls* Vanicek presents characteristics of lower leg amputee gait and then relates that to falling risk. She then presents potential training protocols that may help reduce fall risk.

In their chapters Buckley and Olson examine the identification of and recovery from concussions and low back pain. In *Biomechanical and neuromuscular control factors in determining gait differences in individuals with and without low back pain* Olson presents potential causes of low back pain and how gait affects and is affected by its presence. Buckley reviews current testing for and management of concussion and presents a case that gait, while not currently included in the evaluation of the presence of and recovery from a concussion, should be.

Zhang and Li examine balance control from a central and peripheral perspective while stationary and in walking in their chapter *Feed forward and feedback control for gait and balance*. Initially, they review the neuromuscular system and how it helps to control an inherently unstable system. Then they discuss the fundamentals of balance control including popular theories of standing and gait control. Finally, they review what happens when the system is challenged by weakness and injury.

A novel use of gait is discussed in Hamel and colleagues presentation of *Biometric encryption for gait recognition.* Biometrics is the unique identification of a person based upon specific characteristics. A person's gait may uniquely identify them and be used to help control access or in surveillance.

In: Gait Biometrics
Editors: Li Li and Matthew Holmes
ISBN: 978-1-62948-381-8

Chapter II

A Review of Developments in Gait Asymmetry Quantification

Timothy A. Exell*
School of Electronics and Computer Science, University of Southampton, UK

Abstract

Bilateral asymmetry in gait is a topic that has important implications on injury, performance and data collection. Asymmetry research studies have investigated various forms of gait ranging from sub-maximal velocity walking to maximal velocity sprint running. There have been several recent advancements in the methods used to quantify asymmetry. This chapter critically discusses some of the main developments in asymmetry calculation.

Initial asymmetry studies presented comparable values for left and right sides of the body, but did not calculate explicit measures of asymmetry magnitude. Methods of quantifying asymmetry were developed to meet the need for a dimensionless measure of asymmetry that could be used to compare asymmetry for different discrete variables. Early asymmetry measures, such as the Ratio Index and Symmetry Index, were based on the ratio of left and right values, which could be normalised by a common value. Whilst these methods provided a dimensionless quantification of asymmetry, it has been noted that they can lead to the under or over-inflation of results and can be difficult to interpret. To overcome the problem of over-inflated values, more recent methods have been proposed as alternative measures of gait asymmetry. Recent methods such as the Symmetry Angle provide values that are truly limited to ±100%, overcoming the problem of artificial inflation. However, a limitation associated with the Symmetry Angle is the failure to included intra-limb variability within the asymmetry measure. Since the method requires mean left and right values as input variables, large standard deviations around these mean values may reduce the significance of the reported asymmetry. Recent work has focussed on developing the Symmetry Angle along with other methods used to calculate asymmetry over complete gait cycles to provide a dimensionless asymmetry measure that also incorporates intra-limb variability in the quantification of asymmetry.

* Email: te1@ecs.soton.ac.uk.

The link between kinematic and kinetic asymmetry has been investigated in lower-limb amputees, revealing important asymmetry interactions between the different joints. Recent work has also focused on investigating the relationships between associated kinematic and kinetic variables for individual participants. This work has suggested that compensatory kinetic mechanisms may exist to reduce asymmetry in the kinematic outcome variables. Further work investigating asymmetry of amputee runners and the asymmetry interaction between key gait variables would increase understanding of this topic.

Introduction

Bilateral asymmetry in gait is a topic that has important implications on injury, performance and data collection. Asymmetry has been used in clinical gait settings to quantify inter-limb discrepancies and in running to highlight increased injury potential of one limb over the other (Jacobs et al., 2005). From a performance perspective, knowledge of asymmetry may inform the nature of athletes' training and identify functional asymmetries present for individual athletes, based on performance differences between the two sides of the body (Vagenas & Hoshizaki, 1991). Information about asymmetry in gait also has implications for biomechanical research. Many biomechanical studies of running gait have collected spatio-temporal data unilaterally due to constraints on data collection, such as the positioning of cameras or scanners (Mann & Herman, 1985; Sadeghi et al., 2000; Bezodis et al., 2008; Gittoes & Wilson, 2010). In the event of a large amount of asymmetry being present for an athlete during running, a unilateral analysis could provide an incomplete description of technique and important kinematic and kinetic factors could be overlooked if they occurred in the limb that was not chosen for analysis.

Asymmetry research has explored various forms of gait, from sub-maximal velocity walking to maximal velocity sprint running. Recent years have seen several important developments in the methods used to quantify asymmetry to meet the different analysis demands of discrete and continuous gait data and to incorporate intra-limb variability. This chapter discusses some of the main developments in asymmetry calculation methods and suggests important areas for future research.

The terms symmetry and asymmetry have been used interchangeably in the literature, with symmetry referring to the exact replication of one limb's movement in that of the other. The term asymmetry therefore relates to the amount of divergence of two values from being symmetrical.

Sadeghi et al., (2000) extensively reviewed previous studies into asymmetry and limb dominance in able-bodied gait, highlighting the tendency for symmetry to be assumed during gait analyses. However, since their review, there have been a number of important developments in the methods used to calculate asymmetry during gait. This review will critically discuss the methods available to calculate asymmetry and aims to summarise the key developments in asymmetry calculation. This review is formed of two main sections; the first focussing on the development of asymmetry measures and the second suggesting future avenues for asymmetry research.

Development of Asymmetry Measures

Methods used to calculate asymmetry in gait have developed over time to meet specific researcher's needs. Initial asymmetry quantification relied on simple analyses of discrete data.

Ratio Index

A very basic method on quantifying asymmetry of discrete data is the Ratio Index (Ganguli et al., 1974; Seliktar & Mizrahi, 1986; Andres & Stimmel, 1990; Potdevin et al., 2007). The ratio index is calculated as the ratio of left and right values:

$$R = \frac{x_{left}}{x_{right}} \quad (1)$$

R = Ratio Index

X_{left} = value for left side

X_{right} = value for right side

However, there are numerous limitations associated with this method. One limitation of the Ratio Index relates to the interpretation of the values produced. Using this method, there are no upper limits to the resulting asymmetry values, with a value of 1 indicating perfect symmetry. Another major limitation of the method is the difference in asymmetry results that can occur based on whether the left or right value is larger. If left and right values share the same polarity, and if the left value is smaller than the right value, the Ratio Index will be limited between 0 and 1; however if the left value is larger than the right value, the Ratio Index can be any positive value larger than 1. These limitations make Ratio Index results both difficult to interpret and liable to over or under inflation (Sadeghi et al., 2000). These fundamental limitations to the Ratio Index method led to a need for a more robust asymmetry score.

Symmetry Index

To overcome the limitations associated with the Ratio Index, Robinson et al., (1987) compared force values for left and right legs to determine the effects of chiropractic manipulation on asymmetry. The authors sought a dimensionless quantification of the difference between sides, which would allow asymmetry of different variables to be compared. Robinson et al., (1987) developed the Symmetry Index, a ratio of left and right values divided by a reference value, which was the mean of the two values:

$$SI = \left(\frac{2(X_{left} - X_{right})}{(X_{left} + X_{right})}\right) \times 100\% \quad (2)$$

SI = symmetry index

X_{left} = value for left side

X_{right} = value for right side

Using the Symmetry Index, a value of zero represents absolute symmetry and any divergence from zero indicates asymmetry. The direction of asymmetry is also shown, with a positive Symmetry Index value indicating that the left value was larger than the right value and a negative Symmetry Index value indicating the opposite.

The Symmetry Index method overcame the limitations associated with the Ratio Index. However, Herzog et al., (1989) identified another major limitation with the Symmetry Index method, being that the values produced were not limited to ±100%. In their study, Herzog et al., (1989) collected bilateral force plate data during walking gait and calculated Symmetry Index values for selected kinetic and temporal variables. The authors reported Symmetry Index values in excess of 100% for temporal values and maximum impulse Symmetry Index values were >13 000%. The reason for the inflated asymmetry results was that the left and right values for those variables were either close to zero or positive for one limb and negative for the other limb, which led to a very small reference value that the difference between limbs was divided by. Despite the potential over inflation of Symmetry Index values and results not being limited to ±100%, the method has been very popular in gait asymmetry research (Liu et al., 1998; Astrom & Stenstrom, 2004; Marigold & Eng, 2006; Teichtahl et al., 2009; Kong et al., 2011).

Overall Asymmetry Score

Vagenas and Hoshizaki (1992) proposed another method of calculating asymmetry that included the possibility of an overall asymmetry score for each participant. Initially, asymmetry values were calculated for each variable using a similar method to the Symmetry Index, but using the maximum of the two input values as a reference in place of the mean of both input values:

$$Ia = \left(\frac{(X_{left} - X_{right})}{max(X_{left}, X_{right})}\right) \times 100\% \qquad (3)$$

Ia = Individual asymmetry

X_{left} = value for left side

X_{right} = value for right side

Asymmetry values for each variable were then trichotomised to values of +1, 0 and -1 depending on Ia, so that values ≥1% were 1, values ≤1% were -1 and values in between ±1 were 0. Trichotomised asymmetry values were then summed for all variables to produce an overall asymmetry score for each participant. The ordinal values produced to do not suffer from the artificial inflation associated with the Symmetry Index. However, Sadeghi et al., (2000) noted other limitations of the method, being that asymmetry values would be small, the method could have a low level of sensitivity to small asymmetries and that the location of asymmetry for each participant could not be determined. An additional limitation of the method is that a participant may display large amounts of asymmetry for a number of variables but the direction of asymmetry may not be consistent, which would result in a low overall asymmetry score. For example if a participant displayed larger values on the left side for three variables and larger values on the right side for three different variables, their overall asymmetry score would be zero and would mask the asymmetry displayed by those six variables.

Waveform Method

Crenshaw and Richards (2006) noted that the previous asymmetry measures had been based on discrete values and sought a method of calculating asymmetry for continuous data from a complete gait cycle. Their suggested method compared waveform data over an entire gait cycle with asymmetry calculated based on four measures: trend asymmetry, phase shift, range amplitude ratio and range offset. Trend asymmetry was calculated using eigenvectors to compare time-normalised left and right waveforms (see Crenshaw & Richards, 2006 for details). Phase shift was determined by calculating trend asymmetry and then shifting one of the time-normalised waveforms in 1% increments until asymmetry was minimised. Range amplitude ratio was calculated as the ratio of the range of motion for the left and right limbs. Range offset was calculated as the mean value for the left waveform minus the mean value for the right waveform. Results of the Waveform Method presented by Crenshaw and Richards (2006) were analysed as four individual measures, as opposed to calculating a composite asymmetry score for each participant.

Whilst waveform asymmetry analysis is useful for some variables, such as joint angle profiles, it is also necessary to be able to compare discreet data points. Peak joint angles and forces and impulse values have been identified in previous literature as being highly influential during gait (Mero & Komi, 1994; Weyand et al., 2000; Lelas et al., 2003) and it is, therefore, important that they are included in asymmetry analyses.

Continuous Relative Phase Asymmetry

Another method of calculating waveform asymmetry was proposed by Haddad et al., (2006), who utilised continuous relative phase (see Kelso, 1995) to quantify asymmetry during a complete gait cycle. In their study, Haddad et al., (2006) calculated continuous relative phase between left and right segments during a complete gait cycle. Continuous relative phase analyses have the advantage of providing higher order comparisons between segments that include phase relationships (Varlet and Richardson, 2011). Haddad et al., (2010) compared the calculation of asymmetry across a gait cycle using continuous relative phase with the absolute difference between left and right joint angles and angular velocities. Haddad et al., (2010) reported that continuous relative phase provided a more sensitive measure of asymmetry than the absolute differences for joint angles and angular velocities. Continuous relative phase may provide a useful and sensitive method of calculating higher-order asymmetry over complete gait cycles. However, limitations of this method are that it cannot be used for discrete values at key instances during the gait cycle and that it does not provide information on the specific variables that cause the reported asymmetry.

Symmetry Angle

Zifchock et al., (2008b) addressed the inherent problems associated with the Symmetry Index as a discrete measure of asymmetry, such as the inconsistency in reference values that can be used (Herzog et al., 1989; Vagenas & Hoshizaki, 1992) and how the selection of mean reference values can drastically over-inflate reported findings (Herzog et al., 1989). Zifchock et al., (2008b) proposed another alternative to the SI which, unlike the methods of Crenshaw and Richards (2006) and Haddad et al., (2006), could be used to compare discreet values. They termed their new method the Symmetry Angle (θ_{SYM}, Fig. 1). The θ_{SYM} is an arctan function of the ratio of two contralateral values, where a θ_{SYM} value of 0% indicates perfect symmetry and 100% indicates perfect asymmetry, where values are equal in magnitude and opposite in polarity.

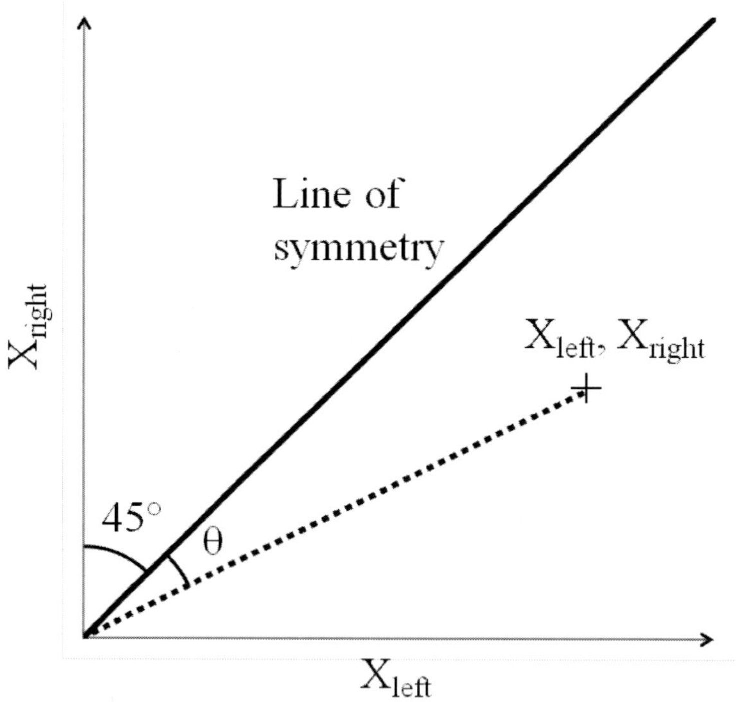

Figure 1. Definition of the Symmetry Angle. Asymmetry is represented by any deviation (θ) from the line of symmetry (adapted from Zifchock et al., 2008b).

Zifchock et al., (2008b) presented two equations for the θ_{SYM} to ensure that values did not exceed 100%. Asymmetry was calculated as:

$$\theta_{SYM} = \frac{\left(45^\circ - arctan(X_{left}/X_{right})\right)}{90^\circ} \times 100\% \tag{4}$$

θ_{SYM} = symmetry angle

X_{left} = value for left side

X_{right} = value for right side

however, if:

$$\left(45^\circ - arctan(X_{left}/X_{right})\right) > 90^\circ \tag{5}$$

then:

$$\theta_{SYM} = \frac{\left(45^\circ - arctan(X_{left}/X_{right}) - 180^\circ\right)}{90^\circ} \times 100\% \tag{6}$$

Zifchock et al., (2008a) highlighted further potential errors with the Symmetry Index, if other reference values were chosen to overcome the artificial inflation caused when using the mean of left and right values. The authors (Zifchock et al., 2008a) compared asymmetry values when using left and right leg values as reference values for Symmetry Index calculation. They reported that across 52 participants, for a range of kinematic and kinetic variables, the majority of participants showed a difference of at least 15% when the opposing leg was used as the reference value. To validate the use of θ_{SYM} as a suitable alternative to the Symmetry Index, the authors assessed the correlation between θ_{SYM} and Symmetry Index values, reporting perfect correlations ($r = 1.00$) for kinetic, structure and strength variables. Correlations for the two kinematic variables were lower initially ($r = 0.63$ and 0.58), however it was suggested by the authors that this was due to some extreme outliers caused by artificially inflated Symmetry Index values, once these outliers were excluded from analysis the correlations increased to $r = 0.78$ and 0.94. It therefore seems reasonable to accept the use of θ_{SYM} as an alternative to Symmetry Index and there are clear advantages to the use of this method as it does not suffer from artificially inflated values.

Consecutive vs Non-Consecutive Foot Contacts

Following the development of the θ_{SYM}, Zifchock and Davis (2008) investigated whether non-consecutive step data could be used for asymmetry analyses. The authors compared θ_{SYM} values for consecutive and non-consecutive footstrikes of 52 male and female runners. They compared both kinematic and kinetic data, reporting a mean difference of just 1.8% between θ_{SYM} calculated from consecutive and non-consecutive footstrikes. To justify the collection of non-consecutive footstrikes, the authors reported CV values both 'between' and 'within' sides. They found that for all variables measured, except for knee adduction, variability between sides was significantly greater than variability within sides. It was noted that the knee adduction results were due to three participants that displayed high levels of variability within a side and that excluding these data sets resulted in the knee adduction results following the same trend as the other variables. Zifchock and Davis (2008) concluded that it seemed acceptable to calculate asymmetry from non-consecutive footstrikes but that care should be taken when high levels of within-side variability are present. These findings reinforce the importance of including intra-limb variability in asymmetry calculations (Giakas & Baltzopoulos, 1997; Exell et al., 2012b). The use of non-consecutive step data for asymmetry analyses can overcome some of the difficulties associated with bilateral data collection, such as restricted data collection volumes (Sadeghi et al., 2000).

Composite Asymmetry Scores

Recent work by Exell et al., (2011; 2012a; 2012b) has built on the previous advances made in asymmetry calculation. The authors adapted the methods presented by Vagenas and Hoshizaki (1992), Crenshaw and Richards (2006) and Zifchock et al., (2008b), proposing customisable composite asymmetry scores that could include both discrete and continuous data (Exell et al., 2012a). An important addition to the composite asymmetry scores presented by Exell et al., (2012a) was the inclusion of intra-limb variability (Giakas & Baltzopoulos, 1997) in determining the significance of asymmetry for each variable. This factor was included in the asymmetry measure by testing for significant differences between left and right values before calculating asymmetry. Asymmetry was then calculated as the positive magnitude of the Symmetry Angle method, multiplied by 0 or 1 based on the significance of the asymmetry:

$$asym(x_n) = (sig) \times ABS\theta_{SYM}(x_n) \tag{7}$$

$asym(x_n)$ = asymmetry value for variable 'x_n'

sig = either 0 or 1 depending on the result of the test of significant difference, with 1 indicating a significant difference between values for the left and right side

$ABS\theta_{SYM}(x_n)$ = magnitude of symmetry angle for variable 'x_n'

The composite scores presented by Exell et al., (2012a) also included the option for an additional test of asymmetry significance to incorporate the effects of asymmetry in the performance outcome, velocity. Individual asymmetry values for specific variables could then be summed to give a composite asymmetry score for each participant, incorporating numerous variables. Methods were also presented to allow for asymmetry of continuous gait data to be incorporated in the composite score. For more details on the composite asymmetry scores see Exell et al., (2012a).

Previous asymmetry scores had neglected to include intra-limb variability, using just the mean left and right values for asymmetry calculation, which could result in deceptively high asymmetry values that are not significant if large standard deviations are present in the values for left or right sides (Exell et al., 2012b). Furthermore, Exell et al., (2012a) highlighted the importance of including variables in asymmetry analyses that are relevant to the analysis being undertaken. Using the composite asymmetry scores, Exell et al., (2011) investigated the interaction between kinematic and kinetic asymmetry for individual athletes during maximal velocity sprint running. Analyses of the asymmetry interactions indicated that a strong link existed between related kinematic and kinetic variables for some but not all athletes, indicating that asymmetry may be used as a compensatory mechanism to reduce the effects of asymmetry on the performance outcome, as previously suggested by Vagenas and Hoshizaki (1991).

Future Directions

Gait asymmetry has been investigated during a range of activities, from walking to maximal velocity sprint running. Studies have indicated the importance of asymmetry on the injury potential of specific limbs of participants; however, most studies provide athlete profiles over a very short time period. Future work would benefit from collecting longitudinal asymmetry data from athletes to establish in greater detail the links between asymmetry, injury and performance. Asymmetry information of this nature would help to inform the coaching-biomechanics interface (Kerwin & Irwin, 2008) and may help to improve running gait from both performance and injury perspectives.

Vagenas and Hoshizaki (1991) and Exell et al., (2011) identified the possibility of functional asymmetries that may exist to reduce the effect of kinetic asymmetry on the performance outcome, and which may be used as a mechanical tool to reduce the injury potential of a particular limb. Further work investigating the phenomenon of functional asymmetries would greatly increase understanding of this area. A mathematical modelling approach may also be beneficial in further investigating the mechanical role of functional asymmetries.

Finally, an area that would benefit greatly from further asymmetry research is that of amputee running gait. Analyses of asymmetry have often formed part of investigations of unilateral amputees during running (Buckley, 1999; Burkett et al., 2003; Brüggemann et al., 2008; Wilson et al., 2009). The inclusion of asymmetry in such analyses is common due to the physiological asymmetry of such athletes. Recently, much attention has been given to the analysis of amputee sprinting due to controversy surrounding the eligibility of amputee athletes to compete in able-bodied competition. Two investigations that arose from the debate surrounding the eligibility of amputee sprinters in able-bodied competition were those of Brüggemann et al., (2008) and Weyand et al., (2009). These two studies assessed different factors relating to the sprint performance of an international bilateral trans-tibial amputee athlete. Due to the differing mechanical demands of intact and prosthetic limbs during sprinting (Weyand et al., 2009), unilateral amputees may have to adapt to the different mechanical demands of one limb whilst maintaining a different technique for the intact limb, whilst a bilateral amputee can focus solely on optimising a symmetrical sprinting technique. Asymmetry quantification may be very useful for both prosthetic design, classification of amputee athletes and mechanical gait analysis of both unilateral and bilateral amputees.

Conclusion

There have been many recent developments in the methods used to calculate gait asymmetry. Methods have been presented so that asymmetry may be calculated for both discrete and continuous data. Following the inclusion of intra-limb variability in the calculation of asymmetry, conclusions can be drawn from experimental data with more confidence than before. Asymmetry has been identified in all forms of gait, ranging from sub-maximal walking to maximal velocity sprint running; however, there is little information available about asymmetry during sprint running and long-term gait asymmetry developments of individual participants. Further research in this area will be beneficial from performance, injury and technology perspectives.

References

Andres, R. O. and Stimmel, S.K. (1990). Prosthetic alignment effects on gait symmetry: A case study. *Clinical Biomechanics*, 5, 88–96.

Astrom, I. and Stenstrom, A. (2004). Effect on gait and socket comfort in unilateral trans-tibial amputees after exchange to a polyurethane concept. *Prosthetics and Orthotics International*, 28(1), 28-36.

Bezodis, I. N., Kerwin, D. G. and Salo, A. I. T. (2008). Lower-limb mechanics during the support phase of maximum-velocity sprint running. *Medicine and Science in Sports and Exercise*, 40(4), 707-715.

Brüggemann, G. P., Arampatzis, A., Emrich, F. and Potthast, W. (2008). Biomechanics of double transtibial amputee sprinting using dedicated sprinting prostheses. *Sports Technology*, 1(4-5), 220-227.

Buckley, J. G. (1999). Sprint kinematics of athletes with lower-limb amputations. *Archives of Physical Medicine and Rehabilitation*, 80(5), 501-508.

Burkett, B., Smeathers, J. and Barker, T. (2003). Walking and running inter-limb asymmetry for Paralympic trans-femoral amputees, a biomechanical analysis. *Prosthetics and Orthotics International*, 27(1), 36-47.

Crenshaw, S. J. and Richards, J. G. (2006). A method for analyzing joint symmetry and normalcy, with an application to analyzing gait. *Gait and Posture*, 24(4), 515-521.

Exell, T. A., Kerwin, D. G., Irwin, G., Gittoes, M. J. R. (2011). Lower-limb biomechanical asymmetry in maximal velocity sprint running. In: Vilas-Boas, J.P., Machado, L., Wangdo, K., Veloso, A.P. (Eds.), Biomechanics in Sports, 29. Portuguese Journal of Sport Sciences 11 (Suppl. 2), 491–494.

Exell, T. A., Gittoes, M. J. R., Irwin, G. and Kerwin, D. G. (2012a). Gait asymmetry: Composite scores for mechanical analyses of sprint running. *Journal of Biomechanics*, 45(6), 1108-1111.

Exell, T. A., Irwin, G., Gittoes, M. J. R. and Kerwin, D. (2012b). Implications of intra-limb variability on asymmetry analyses. *Journal of Sports Sciences*, 30(4), 403-409.

Ganguli, S., Mukherji, P. and Bose, K. S. (1974). Gait evaluation of unilateral below-knee amputees fitted with patellar-tendon-bearing prostheses. *Journal of the Indian Medical Association*, 63(8), 256–259.

Giakas, G. and Baltzopoulos, V. (1997). Time and frequency domain analysis of ground reaction forces during walking: an investigation of variability and symmetry. *Gait and Posture*, 5(3), 189-197.

Gittoes, M. J. R. and Wilson, C. (2010). Intralimb joint coordination patterns of the lower extremity in maximal velocity phase sprint running. *Journal of Applied Biomechanics,* 26(2), 188-195.

Haddad, J. M., Van Emmerik, R. E. A., Whittlesey, S. N. and Hamill, J. (2006). Adaptations in interlimb and intralimb coordination to asymmetrical loading in human walking. *Gait & Posture,* 23, 429–434.

Haddad, J. M., Van Emmerik, R. E. A., Wheat, J. S., Hamill, J. and Snapp-Childs, W. (2010). Relative phase coordination analysis in the assessment of dynamic gait symmetry. *Journal of Applied Biomechanics,* 26, 109–113.

Herzog, W., Nigg, B. M., Read, L. J. and Olsson, E. (1989). Asymmetries in ground reaction force patterns in normal human gait. *Medicine and Science in Sports and Exercise,* 21(1), 110-114.

Jacobs, C., Uhl, T. L., Seeley, M., Sterling, W. and Goodrich, L. (2005). Strength and fatigability of the dominant and nondominant hip abductors. *Journal of Athletic Training*, 40(3), 203-206.

Kerwin, D. G. and Irwin, G. (2008). Biomechanics for coaches. In R. L. Jones, M. Hughes and K. Kingston (Eds.), *An introduction to sports coaching* (pp 87-100). London, UK, Routledge.

Kelso, J. A. S. (1995). *Dynamics Patterns: The Self-Organization of Brain and Behavior.* MIT Press, Cambridge.

Kong P. W., Candelaria N.G. and Smith, D. (2011). Comparison of longitudinal biomechanical adaptation to shoe degradation between the dominant and non-dominant legs during running. *Human Movement Science*, 30(3), 606-613.

Lelas, J. L., Merriman, G.J., Riley, P.O. and Kerrigan, D.C. (2003). Predicting peak kinematic and kinetic parameters from gait speed. *Gait and Posture*, 17(2), 106-112.

Liu, X. C., Fabry, G., Molenaers, G., Lammens, J. and Moens, P. (1998). Kinematic and kinetic asymmetry in patients with leg-length discrepancy. *Journal of Pediatric Orthopaedics*, 18(2), 187-189.

Mann, R. and Herman, J. (1985). Kinematic analysis of Olympic sprint performance: Men's 200 meters. *International Journal of Sport Biomechanics,* 1, 151-162.

Marigold, D. S. and Eng, J. J. (2006). The relationship of asymmetric weight-bearing with postural sway and visual reliance in stroke. *Gait and Posture,* 23(2), 249-255.

Mero, A. and Komi, P. V. (1994). Emg, force, and power analysis of sprint-specific strength exercises. *Journal of Applied Biomechanics,* 10(1), 1-13

Potdevin, F. J., Femery, V. G., Decatoire, A., Bosquet, L., Coello, Y. and Moretto, P. (2007). Using effect size to quantify platar pressure asymmetry of gait of nondisabled adults and patients with hemiparesis. *Journal of Rehabilitation Research and Development*, 44(3), 347-354.

Robinson, R. O., Herzog, W. and Nigg, B. M. (1987). Use of force platform variables to quantify the effects of chiropractic manipulation on gait symmetry. *Journal of Manipulative Physiological Therapy,* 10(4), 172-176.

Sadeghi, H., Allard, P. and Duhaime, M. (2000). Contributions of lower-limb muscle power in people without impairments. *Physical Therapy,* 80(12), 1188-1196.

Seliktar, R. and Mizrahi, J. (1986). Some gait characteristics of below knee amputees and their reflection on the ground reaction forces. *Engineering in Medicine*, 15(1), 27–34.

Teichtahl A. J., Wluka A. E., Cicuttini F. M., Davis S. R. and Morris M. E. (2009). The Associations Between the Dominant and Nondominant Peak External Knee Adductor Moments During Gait in Healthy Subjects: Evidence for Symmetry. *Archives of Physical Medicine and Rehabilitation*, 90(2), 320-324.

Vagenas, G. and Hoshizaki, B. (1991). Functional asymmetries and lateral dominance in the lower limbs of distance runners. *International Journal of Sports Biomechanics,* 7(4), 311-329.

Vagenas, G. and Hoshizaki, B. (1992). A multivariable analysis of lower extremity kinematic asymmetry in running. *International Journal of Sports Biomechanics*, 8(1), 11-29.

Varlet, M. and Richardson, M.J. (2011). Computation of continuous relative phase and modulation of frequency of human movement. *Journal of Biomechanics*, 44, 1200-1204.

Weyand, P. G., Bundle, M. W., McGowan, C. P., Grabowski, A., Brown, M. B., Kram, R. and Herr, H. (2009). The fastest runner on artificial legs: different limbs, similar function? *Journal of Applied Physiology*, 107(3), 903-911.

Weyand, P. G., Sternlight, D. B., Bellizzi, M. J. and Wright, S. (2000). Faster top running speeds are achieved with greater ground forces not more rapid leg movements. *Journal of Applied Physiology,* 89(5), 1991-1999.

Wilson, J. R., Asfour, S., Abdelrahman, K. Z. and Gailey, R. (2009). A new methodology to measure the running biomechanics of amputees. *Prosthetics and Orthotics International*, 33(3), 218-229.

Zifchock, R. A. and Davis, I. (2008). Non-consecutive versus consecutive footstrikes as an equivalent method of assessing gait asymmetry. *Journal of Biomechanics,* 41(1), 226-230.

Zifchock, R. A., Davis, I., Higginson, J., McCaw, S. and Royer, T. (2008a). Side to side differences in overuse running injury susceptibility: A retrospective study. *Human Movement Science*, 27(6), 888-907.

Zifchock, R. A., Davis, I., Higginson, J. and Royer, T. (2008b). The symmetry angle: a novel robust method of quantifying asymmetry. *Gait and Posture*, 27(4), 622-627.

In: Gait Biometrics
Editors: Li Li and Matthew Holmes

ISBN: 978-1-62948-381-8

Chapter III

Patterned Variability in Gait Behavior: How Can it be Measured and What Does it Mean?

***Christopher K. Rhea*[1,*] *and Adam W. Kiefer*[2,3]**
[1]Department of Kinesiology, University of North Carolina at Greensboro, Greensboro, NC, US
[2]Division of Sports Medicine, Cincinnati Children's Hospital Medical Center, Cincinnati, OH, US
[3]Department of Pediatrics, College of Medicine, University of Cincinnati, Cincinnati, OH, US

Abstract

Humans do not exhibit consistent behavior when walking. Every successive stride is slightly different and the magnitude of stride variation has traditionally been used to identify gait control dysfunction. Over the past two decades, however, the manner in which gait variability is quantified, and ultimately interpreted, has experienced a paradigm shift as researchers have begun to exploit a theoretical framework based on nonlinear dynamics. This has expanded the focus beyond traditional metrics and toward an examination of stride variability dynamics, and has led to further insight into how humans adapt their gait to prevent slips and trips, and avoid collisions. This chapter first examines the variable nature of gait, how constraints influence gait variability, and expounds upon the theoretical framework of dynamical disease. Next, four classes of commonly used computational methods within the gait variability literature are discussed. The chapter concludes with a discussion of the implications this approach has for theoretical and clinical research, as well as what such a framework affords researchers in the future.

* Corresponding author: Christopher K. Rhea, E-mail: ckrhea@uncg.edu.

Introduction

Despite over a century of research on human gait and the underlying mechanisms of locomotor control, the capability of humans to adaptively walk in a dynamic, complex world is still not well understood.

This is because adaptive, functional gait emerges from the complex interactions within and across physiological, neuromuscular and behavioral scales (Diniz et al., 2011). At one scale, the functionality of a biological system can be judged on its ability to exhibit repeatable (i.e., stable) behavior.

For example, in the context of physiology, and more specifically the human cardiovascular system, it is desirable to have a consistent heart rate while at rest. However, the system also has to remain appropriately flexible so as to increase heart rate to meet physiological demands as the individual transitions through different states of behavior (e.g., the progression from a low-frequency to a high-frequency state of locomotion).

Over the last 30 years, the theoretical framework of *dynamical disease*—a framework rooted in dynamical systems theory—has allowed for much development toward the understanding of functional behavior in physiological systems (Belair, Glass, Heiden, and Milton, 1995; Glass and Mackey, 1979; Goldberger, Rigney, and West, 1990; Hausdorff, Peng, Ladin, Wei, and Goldberger, 1995; Mackey and Glass, 1977; Mackey and Milton, 1987; Peng et al., 1993; West and Goldberger, 1987). Proponents of this theory assert that a change in the repeatability of the dynamics of a healthy physiological system (e.g., the structure of variability of the inter-beat intervals of the human cardiac rhythm) that arise due to pathology, for example, are informative about the adaptive nature of the system. This approach has been used to examine a variety of physiological control systems, including respiration, blood chemistry, electrophysiology, and hormone regulation (Mackey and Milton, 1987). A central tenet of this approach is that the amount of variability exhibited by a system is not as important as the way in which the variability is temporally structured, and it is this that determines whether such a system is functional (i.e., adaptive) or not (i.e., maladaptive).

More recently, the dynamical disease approach has been expanded in an effort to describe the functionality of healthy and pathological locomotor systems in a variety of different experimental contexts (Dingwell and Cusumano, 2000; Hausdorff, 2007; Hausdorff, et al., 1995; Jordan, Challis, Cusumano, and Newell, 2009; Rhea, Kiefer, Gerin-Lajoie, Warren, and Aaron, under review; Scafetta, Marchi, and West, 2009; Stergiou and Decker, 2011; Vaillancourt and Newell, 2002).

The main benefit of this approach to the study of human gait is that the neuromuscular system can be considered in much the same way as a purely physiological control system. That is, its adaptive capabilities are shaped by the constraints on the actor (e.g., pathology), the task (e.g., difficulty of a movement pattern), and the environment (e.g., ground surface changes) (Newell, 1986). Given the nature of these constraints, and their interaction across many different time scales, a dynamical disease approach requires the utilization of computational methods beyond classical approaches.

This chapter is dedicated to exploring the use of these methods to examine the role of variability in human gait. Specifically, we will introduce several computational methods that are commonly used to quantify gait dynamics, discuss how gait dynamics are altered by constraints, and offer some solutions for the integration of the dynamical disease framework into clinical rehabilitation.

However, before we move further into a discussion of these computational methods it is first necessary to understand the nature of variability in human gait and the role it plays in the adaptability of human locomotion.

Variable Nature of Gait

The next time you are out on a walk somewhere, take a look at the world around you. It is filled with irregular terrain, such as steps, curbs, and potholes that you must routinely traverse. There are also stationary and moving objects such as street signs, benches, cars, and pedestrians that must be accounted for to avoid a collision and prevent a potential injury. Thus, gait must be appropriately adapted to successfully maneuver through the environment. This requires repeatable behavior, and specifically the ability to reliably put one foot in front of the other. Humans do not exhibit consistent gait behavior, however, because they also must maintain the ability to change behavior to meet the demands of the environment. In fact, every stride differs slightly from the stride that precedes it and the presence of stride-to-stride variability has been documented for over a century (Vierordt, 1881); although it has only become a point of interest for movement scientists in the last two decades. Classically, stride-to-stride variability (commonly expressed as the standard deviation around the mean) was equated to imprecision in walking mechanics and it was thought to have no functional value. This construct was grounded in traditional motor control theories that used fluctuations in behavior as a performance measure, and these theories were useful in describing motor learning during skill acquisition (Adams, 1971; Schmidt, 1975). Specifically, such theories focused on the *amount of variability* in the movement pattern. This was typically indexed by the calculation of a summary metric of the variation over the entire performance, such as range, standard deviation or coefficient of variation. The focus was for the human to minimize the amount of variability in order to produce more consistent behavior. For example, stride-to-stride variability decreases throughout a child's development, presumably as the child learns to more accurately control his/her gait (Hausdorff, Zemany, Peng, and Goldberger, 1999). The amount of variability has also been used to explain when control is degraded. For example, stride-to-stride variability increases after the onset of Parkinson's disease or Huntington's disease (Hausdorff, Cudkowicz, Firtion, Wei, and Goldberger, 1998). This has led traditional performance and rehabilitation protocols to consider a reduction of gait variability to be indicative of enhanced performance (Olmo and Cudeiro, 2005; Stolze et al., 1997; von Porat et al., 2006). An unintended consequence of this traditional position is that functional behavior can be minimized, from a theoretical perspective, through rigid, consistent behavior. However, this runs antithetical to what is required to navigate the world around us: a world that, given its exceptional complexity, requires human behavior to maintain flexibility and remain poised for transitions between stable behavioral states (Kiefer et al., under review; Kloos and Van Orden, 2010; Van Orden, Kloos, and Wallot, 2011). Even when an individual traverses uniform terrain, every stride is slightly different because the position of the center of mass is constantly changing relative to the previous stride. This requires fine-grained adjustments at short and long time scales, and the ability to adapt the stride to respond to such center of mass fluctuations is a requirement for the emergence of functional gait. The challenge of gait adaptation is amplified when considered in the context of a complex and dynamic environment in which the individual must act. The ability for a gait system to adapt to these constraints is paramount for the existence of functional mobility. The preceding examples provided the foundation for the importance of an alternative framework called *dynamical disease*; an approach that not only examines the utility of gait variability, but makes variability a requirement as a basis for functional mobility.

Furthermore, in order for this framework to continue to gain traction, the traditional approach to how variability is measured and understood has to be challenged. Summary metrics like standard deviation are useful to gain an overarching view of a behavior. However, the locomotor system evolves over time and such measures most often miss the subtle, yet important fluctuations exhibited in the gait dynamics because they average the fine-grained variations across the entire behavior. An alternative way to quantify behavior is to examine the *structure of variability*. This provides a description of how the variability changes over time and across scales, and provides greater insight into the influence of constraints on the control system.

As discussed earlier, traditional motor control theories suggested that movement variability was simply a part of a random process. By 1995, the work by Hausdorff et al., challenged the classic assumption that gait variability was completely random when the presence of long-range correlations was identified in the structure of stride-to-stride variability during human locomotion (Hausdorff, et al., 1995). These long-range correlations provided a measure of self-similar patterns that repeated throughout the behavior across several temporal scales. Thus a pattern of variability observed early in the behavior might repeat again not only in the steps that immediately follow, but also in steps that occurred seconds or even minutes later. This was an important observation and patterns within gait variability have since been suggested to offer functional value (Glass, 2001; Hausdorff, 2007; Lipsitz, 2002; Stergiou and Decker, 2011). We have already discussed that a relatively invariable system lacks functionality due to its inability to adapt to constraints. Hhowever, a variable system does not necessarily equate to functionality. Variability may manifest in a number of ways that might help or hinder performance. For example, the gold standard for functional movement is typically the behavior exhibited by young, healthy adults. This is when the muscular, skeletal, and neurological systems have reached maturity, allowing for optimal gait adaptability. Accordingly, the patterned variability observed in young healthy adults exhibits both repeatability and adaptive variation concurrently. This is ideal for the system to respond to internal and external demands while walking. However, a change to the muscular, skeletal, or neurological systems may lead to an altered structure of gait variability, and this may lead to a decrement in functional mobility. For example, less prominent long-range correlations are observed in older adults (Hausdorff et al., 1997). This is presumably due to a degradation in the neuromuscular system that may relate to the increased fall rate in older adults (Hausdorff, 2007).

Task constraints can also change the patterns observed in gait variability, such as when people walk or run faster or slower than their preferred speed (Jordan, Challis, and Newell, 2006, 2007b). This is because changes to the constraints acting on a physiological system may alter the functionality of the system[1]. Constraints on the actor-task-environment system are a central theme of the extension of dynamical disease into motor control, and the influence of constraints will be explored in a variety of contexts throughout the rest of this chapter.

First, however, it is important to clarify the various methods that have been used to index the patterns in gait dynamics, as these diverse methodologies make it difficult to compare results across studies. Thus, the next section outlines a few methodological issues that should be considered prior to an examination of gait dynamics.

Methodological Considerations

There are two methodological issues that need to be considered prior to indexing the patterns of variability in gait: (1) the measurement technique used to record the data from which the patterns are being examined and (2) the quantification of these patterns. Much of the early work that examined the structure of variability in physiological systems focused on discrete time intervals between measureable events, such as the beat-to-beat intervals of the human cardiac rhythm (Goldberger, 1991; Goldberger, et al., 1990; Goldberger and West, 1987; Ivanov et al., 1999; Kaplan et al., 1991; Lipsitz and Goldberger, 1992; Peng, Havlin, Stanley, and Goldberger, 1995; Peng, et al., 1993; Stanley et al., 1999; West and Goldberger, 1987). Each time interval was recorded across several hundred heartbeats, and this data was used to generate a time series of cardiac behavior. The time series of the beat-to-beat intervals was then examined for repeating patterns.

Similarly, early studies of the dynamics of gait variability analyzed the discrete time intervals between strides (Hausdorff et al., 2000; Hausdorff, et al., 1997; Hausdorff, et al., 1995; Hausdorff et al., 1996; Hausdorff, et al., 1999; West and Griffin, 1999; West and Scafetta, 2003). We term this approach the *interval method*. While this method has proven useful, it is limited given that many of the analyses that we will cover throughout this chapter require several hundred data points to be reliable.

Since only one data point is recorded for each stride when using the interval method, this requires several hundred steps (typically 10-20 minutes of walking) to accurately assess the gait dynamics, and this amount of data may not be possible to obtain if the population of interest exhibits limited mobility. A second way to examine gait patterns is via an approach we term the *continuous method*. This method indexes a continuous variable, such as knee angle throughout the stride, and creates a time series comprised of the gait behavior at every recorded instant.

The number of data points recorded in the continuous method is dependent on the sampling frequency, and this potentially solves the trial length problem. It is important to note, however, that thoughtfulness should be exhibited when this method is utilized because

[1] It should be noted that the early work relating physiological variability to functionality described a change in variability from a patterned state to a random state as a decline in functionality. This model gained popularity as the "loss of complexity hypothesis" [33]. More recent work has described any deviation away from the patterned behavior exhibited by young, healthy adults (either toward less patterned or more patterned) as a decrement in functionality. Less patterned behavior is often times undesirable because it may indicate less fine-tuned control, whereas an increase in patterned behavior can be undesirable because it may represent less adaptability in response to internal and external demands.

gait behavior sampled at high frequencies (e.g., 500 Hz) would yield enough data points required for many of the analyses of patterned behavior after just a few strides. However, it is unlikely that any discernable information about the gait dynamics can be gained by an examination of three or four strides. The analyses discussed later in this chapter index how gait dynamics evolve over time, so an examination of the variability over a small window of time is likely to miss potentially important low-frequency fluctuations of the patterns as they evolve over time. While the interval and continuous methods are prevalent in the literature, there is a lack of consensus as to which method more accurately quantifies patterns in gait behavior. It likely depends on the nature of the research question and potentially, the etiology of a specific pathological population.

The second methodological issue encompasses the way in which gait dynamics are quantified. Many algorithms have been developed over the past two decades (many outside of the field of physiology), and each provides a different way to characterize the structure of variability in a system (Bravi, Longtin, and Seely, 2011). These different computational methods can potentially allow for a convergence of evidence as to how gait dynamics evolve by indexing similar changes in the patterns of behavior through different analyses. Unfortunately, in many cases it is unclear how these analyses relate to each other, and also how each analysis relates to functional motor behavior. The next section is dedicated to introducing some of the most commonly used computational methods to index patterns in the structure of variability in gait behavior with both the interval and continuous methods of measurement.

Computational Methods to Quantify the Structure of Gait Variability

A number of computational methods have been proposed to identify patterns in movement variability (Bravi, et al., 2011). This section provides an overview of four classes of computational methods that have been commonly used to quantify the dynamics of gait variability. Each classification provides a unique way to index patterns, and many of the classes include sub-categories that provide further insight into the dynamic properties of gait. There is no consensus as to which set of computational methods provides the most accurate index of functional mobility, as each method indexes a different component of the dynamic pattern. Accordingly, the classes are presented in no particular order.

Long Range Correlations

Detrended fluctuation analysis (DFA) was first developed to examine the mosaic structure of DNA nucleotides (Peng et al., 1994), and was born from the idea that variation in DNA sequences was not random, but had a particular structure. DFA is an extension from an earlier developed method called fluctuation analysis (FA) (Peng et al., 1992). Many physiological systems exhibit non-stationarity over time which, based on the time scale, can influence how accurately the structure of variability can be quantified (Peng, et al., 1994; Peng, et al., 1995; Peng, et al., 1993). Therefore, the "detrended" portion of DFA was added to FA to more accurately index the dynamic patterns in the face of non-stationarities within the system. Prior to the removal of non-stationarity (i.e., detrending the data), the time series is first integrated by subtracting the mean from every data point through the following equation:

$$y(k) = \sum_{i=1}^{k}[S(i) - S_{ave}], \tag{1}$$

where $S(i)$ is the ith step number and S_{ave} is the average step interval (Figure 1a). The integrated time series $y(k)$ is then portioned into boxes that consist of an equal number of data points, and a trend line is fit to the data contained within each box (Figure 1b).

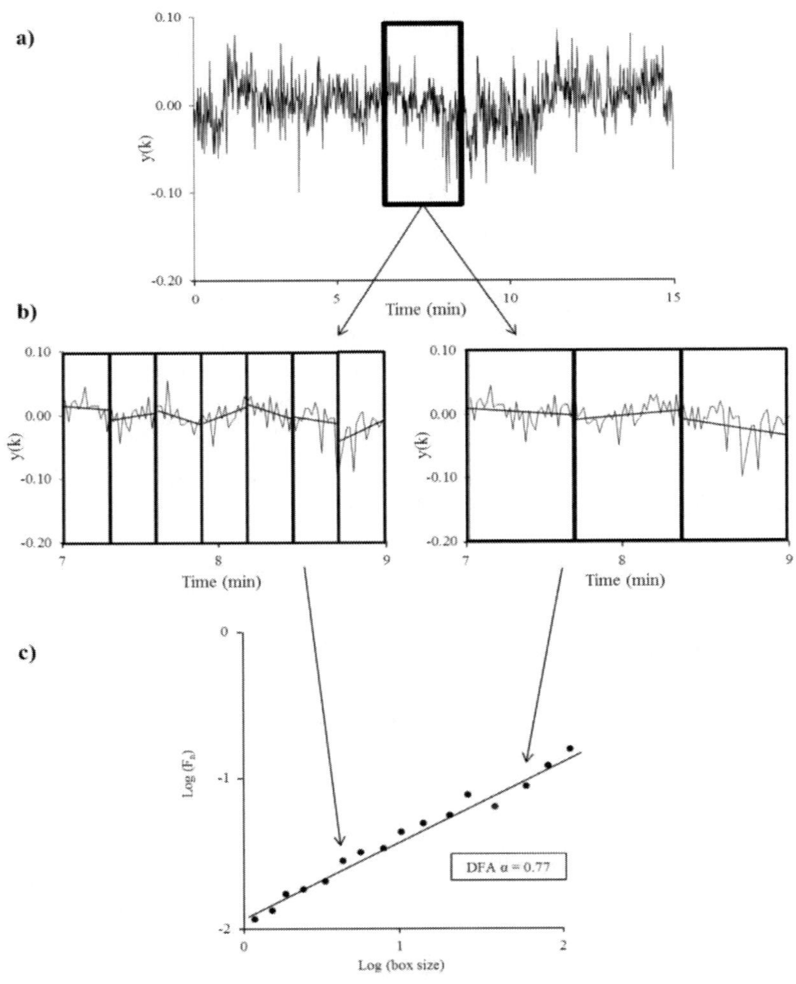

Figure 1. Detrended Fluctuation Analysis (DFA). The stride interval time series is first demeaned using the y(k) function (a). Next, entire time series is partitioned into a series of box sizes that contain an equal number of data points. For visual clarity, a subset of the time series (minutes 7-9) have been zoomed in on and two of the box sizes are shown (b). A trend line is fitted within each box and the residual variability above and below the trend line (effectively detrending the data) is calculated using the F_n function. The residual variability within each box size is then plotted on a log-log plot and a least squares line is fit to the data. The slope of the least squares line is the DFA α metric, with values for healthy gait tending toward 0.75. DFA α for the healthy participant shown is 0.77.

After the local trend is removed via the detrended fluctuation function, the remaining fluctuations are quantified using the root-mean-square method:

$$F(n) = \sqrt{\frac{1}{N}\sum_{k=1}^{N}[y(k) - y_n(k)]^2}. \quad (2)$$

This process is iterated across box sizes of increasing size that typically range from 4 data points long to a maximum length of ¼ of the entire time series length (Duarte and Zatsiorsky, 2001; Kiefer, Riley, Shockley, Villard, and Van Orden, 2009; Roerdink et al., 2006). The root-mean-square of the detrended fluctuations at each box length is then plotted on a log-log plot and a least squares line indexes the slope of the data (Figure 1c). This slope corresponds to the *DFA alpha (α)* value, and is a measure of the strength of the long-range correlations. DFA α typically ranges from 0.5 to 1.0 in human stride intervals (Hausdorff, et al., 1995; Hausdorff, et al., 1996), with values tending toward 0.5 indicating weaker long-range correlations (i.e., more random behavior) and values tending toward 1.0 indicating stronger long-range correlations (i.e., more patterned behavior). As previously stated, functional gait behavior exhibits both repeatability and adaptive variation, so neither overly random behavior nor overly correlated behavior is desirable from a functional perspective. Accordingly, DFA α values typically exhibited by young, healthy adults are ≈ 0.75.

As outlined by the dynamical disease framework, a change to the constraints that influence the control system may correspond to a change in the structure of variability in the dynamics of the system. For example, constraint changes may result from a neurological insult, ultimately influencing the structure of gait variability and reduce functional mobility. Research has demonstrated that weaker long-range correlations are observed in older adults and patients with Huntington's disease, Parkinson's disease, and amyotrophic lateral sclerosis (Hausdorff, et al., 2000; Hausdorff, et al., 1997), presumably due to degradation in neural control. Interestingly, long-range correlations are preserved when the neurological insult is confined to the peripheral nervous system, as seen in patients with peripheral neuropathy (Gates and Dingwell, 2007) and a ruptured anterior cruciate ligament (ACL) (Rhea, Kiefer, D'Andrea, Warren, and Aaron, 2010).

A rationale for this phenomena is that long-range correlations may originate in the central nervous systems, and thus an insult to the peripheral nervous system may not influence the gait dynamics indexed by DFA α (Gates and Dingwell, 2007). However, there is much debate as to the origin of long-range correlations in human behavior (Diniz, et al., 2011), so more research is needed to fully explain the presence of long-range correlations in the gait dynamics following a neurological insult to the peripheral nervous system.

The experimental task can also alter constraints, and researchers have examined this by the modification of treadmill speed (faster or slower than participants' preferred speed) as participants walk or run (Jordan, et al., 2009; Jordan, et al., 2006; Jordan, Challis, and Newell, 2007a; Jordan, et al., 2007b).

Preferred walking speed is characterized by stable coordination patterns in the lower limbs, and a deviation away from preferred walking speed destabilizes these coordination patterns (Diedrich and Warren, 1995). This destabilization may present as either a more random or more rigid pattern, both of which would lead to increased difficulty in appropriate gait adjustment (i.e., maladaptation). For example, DFA α increases when individuals walk or run above or below their preferred speed.

The increase in long-range correlations represents greater repeatability in the gait dynamics, potentially at the cost of adaptability. The fact that constraints can lead to DFA α shifting toward weaker long-range correlations (e.g, after a neurological insult) or toward stronger long-range correlations (e.g., due to a change in task) highlights that an appropriate amount of repeatability and adaptive variation in gait behavior is required for a functional locomotor system. This has led to the development of a number of theoretical models to describe how a bidirectional change from an appropriate structure of variability influences adaptive control (Harbourne and Stergiou, 2009; Stergiou, Harbourne, and Cavanaugh, 2006; Vaillancourt and Newell, 2002).

Entropy

The concept of entropy can be traced back to the second law of thermodynamics, although the term has been used in a variety of contexts (Williams, 1997). Specific to physiological systems, entropy is used to describe the complexity of the system's behavior by indexing the regularity of patterns present in the dynamics. A direct measure of entropy, such as Kolmogorov-Sinai (K-S) entropy, requires a large, noise free data set (Williams, 1997). Since the data from physiological systems are relatively short and never noise free, a number of algorithms have been developed to estimate entropy. Approximate entropy (ApEn) was introduced as a computational method that quantified the number of repetitive patterns in a data set (Pincus, 1991). First, a template is created from a small subset of the time series and then compared to all successive data points to determine if a match to the template is observed. A requirement of this analysis is the determination of the length (*m*) of a template pattern and the level of tolerance (*r*) that is acceptable to be deemed a match to the template. For gait, m=2 and r=0.2 of the standard deviation of the time series have typically been used (Georgoulis, Moraiti, Ristanis, and Stergiou, 2006; Moraiti et al., 2009). Thus, for a stride interval time series S_n, the first two data points in the time series (S_1 and S_2) would define the first template, and the template would be compared to all consecutive two data point patterns within the time series. In this example, if S_1 and S_2 are within $\pm\, r$ of the corresponding value in the pattern, then a match is recorded. Next, a template is created from the subsequent two data points (S_2 and S_3) and the process is repeated. After all pairs of data point templates have been used to index repeating patterns, the ratio $C_m(r)$ is calculated by dividing the total number of observed matches to the templates by the total number of possible observed matches to the templates, and expressed relative to the template length and tolerance value. For example, if 27 template matches were observed out of 100 possible matches using m=2 and r=0.2*SD of S_n, then $C_2(0.2*SD)$ = 27/100 = 0.27. The process is repeated for $m + 1$ using the same *r,* and ApEn is given by the following equation:

$$\text{ApEn}(S_n,m,r) = \ln\left[\frac{C_m(r)}{C_{m+1}(r)}\right]. \tag{3}$$

Output values from ApEn typically range from 0 to 2 in physiological systems (Arif, Ohtaki, Nagatomi, and Inooka, 2004; Buzzi and Ulrich, 2004; Cavanaugh et al., 2006; Georgoulis, et al., 2006; Vaillancourt and Newell, 2000), with values that tend toward 0 indicative of more repeatability (i.e., less complex behavior) and values that tend toward 2 indicative of less repeatability (i.e., more complex behavior). For example, a very repeatable signal such as a sine wave would have values very close to 0 and would be deemed to have low complexity. As the periodic structure of the sine wave breaks down (e.g., noise is introduced into the signal), ApEn values trend away from 0. This indicates more complex behavior and less repeatability in the signal.

Consistent with the dynamical disease framework, ApEn of gait behavior changes when the constraints on the system are altered. For example, a ruptured ACL in the knee induces a neurological insult due to the disruption of proprioceptive information available from the mechanoreceptors in the ruptured ligament, while also decreasing the structural integrity of the joint. One may hypothesize that this type of neurological insult would lead to an increase in the ApEn of gait behavior (i.e., less repeatability) due to the derogated information available for feedback. However, the opposite is reported when gait behavior of a patients' ACL-ruptured limb is compared to their unaffected limb (Georgoulis, et al., 2006) and when patients' affected limbs are compared to the matched limb of healthy controls (Rhea, et al., 2010). This shift toward more consistent gait behavior is likely due to the co-contraction of muscles around the knee acting as a protective mechanism after an ACL rupture to avoid further structural injury (Hurd and Snyder-Mackler, 2007; Lindstrom, Fellander-Tsai, Wredmark, and Henriksson, 2010; Osternig, Caster, and James, 1995; Rudolph, Axe, Buchanan, Scholz, and Snyder-Mackler, 2001).

As a consequence, more consistent gait behavior emerges and this decreases functionality because the ability to adaptively vary locomotor behavior to comply with demands is diminished. Similarly, a decline in functionality can occur when task constraints are altered, such as walking below or above one's preferred speed (Buzzi and Ulrich, 2004; Georgoulis, et al., 2006), as fatigue increases (Arif, Othaki, Nagatomi, Ishihara, and Inooka, 2002) or when individuals walk forward or backward in virtual reality (Katsavelis, Mukherjee, Decker, and Stergiou, 2010a, 2010b).

It should be noted that an enhanced version of ApEn, called sample entropy (SampEn) has been developed (Richman and Moorman, 2000), given the limitations ApEn has with noisy data and self-matching templates. SampEn is shown to be more robust to noise (Ramdani, Bouchara, and Lagarde, 2009; Rhea et al., 2011), especially when a delay is employed (Govindan, Wilsona, Eswaran, Lowery, and Preißl, 2007), and has greater utility with short data sets. Since noise is inherent to any biological system and most researchers are confined to relatively short data sets, some researchers have transitioned to SampEn over ApEn for a more accurate estimation of entropy.

However, most SampEn analyses in the motor behavior literature are limited to the control of posture (Donker, Ledebt, Roerdink, Savelsbergh, and Beek, 2008; Lamoth, Ainsworth, Polomski, and Haudijk, 2010; Perlmutter, Lin, and Makhsous, 2010; Roerdink, et al., 2006). In the future it is inevitable that a greater number of researchers will begin to employ SampEn as opposed to ApEn in their analysis of the dynamics of gait.

Local Dynamic Stability

It is important to note that "stability" has taken on a variety of meanings in the biomechanics literature (Reeves, Narendra, and Cholewicki, 2007). In a traditional sense, stability is defined as the ability to resist perturbations.

In gait, the ability to recover from slips and trips is termed global stability (Dingwell, Cusumano, Sternad, and Cavanagh, 2000; Dingwell and Marin, 2006; Dingwell, Robb, Troy, and Grabiner, 2008) and many clinical measures have been developed to capture a patient's stability level to identify potential fall risk, such as the Berg Balance Score (Berg, Wood-Dauphinee, Williams, and Gayton, 1989), the Dynamic Gait Index (Shumway-Cook and Woollacott, 1995), and the Functional Gait Assessment (Wrisley, Marchetti, Kurharsky, and Whitney, 2004). This section is dedicated to a different type of stability termed local dynamic stability. This construct characterizes the ability of a system to resist small perturbations, and is quantified by the Lyapunov exponent. Local dynamic stability does not always equate to global stability, so the two should not be used interchangeably. Only recently have researchers begun to map out the relation between local dynamic stability and global stability, specifically in the context of predictive models (Bruijn, Bregman, Meijer, Beek, and Dieen, 2012); however, more modeling and experimental work in this area is necessary. Regardless, a lot has been learned about the nature of gait dynamics through the examination of the local dynamic stability of gait behavior in natural and experimentally constrained contexts.

The local dynamic stability of a system is indexed by the examination of a continuous variable (e.g., knee angle time series) as it evolves over time. However, all physiological signals are a composite of the interaction of the sub-components of the system, and all have unique dynamic properties (Diniz, et al., 2011). This makes it difficult to characterize the dynamics of a system within a single dimension. By examining the system's dynamics in multiple dimensions, the interaction of the behavior in different dimensions is more accurately characterized and allows for a more parsimonious description of the dynamics. In multidemsional systems (i.e., a system with multiple dynamic sub-components such as the human locomotor system), the dynamic characteristics that underlie the system can be properly examined by plotting the original time series against time-delayed copies of itself (Takens, 1981). This creates a multidimensional state space for which to examine, and ultimately quantify, patterns of behavior. For example, a time delayed copy of a knee angle time series S_n would be created by the following equation:

$$s_n = [S_n(i), S_n(i + t), S_n(i + 2t), S_n(i + 3t), \ldots, S_n(i + (d_E - 1)t], \tag{4}$$

where s_n is the time delayed time series, S_n is the original time series, t is the time delay and d_E is the embedding dimension. For gait, the three most commonly used methods to select an appropriate t, which is dependent on the dynamic nature of the data, are the autocorrelation (King, Jones, and Broomhead, 1987) average mutual information (AMI) (Fraser and Swinney, 1986) and average displacement (AVD) (Rosenstein, Collins, and Luca, 1994) techniques. While autocorrelation and AMI are still abundantly used, AVD has demonstrated to be more robust with regard to noisy signals and should be considered an appropriate method for the selection of t in physiological time series. The false nearest neighbors (FNN) technique can be employed to determine the correct d_E needed to adequately unfold the dynamics (Kennel, Brown, and Abarbanel, 1992). Once d_E is determined, the creation of a time-delayed copy is iterated until all of the desired time series have been created (one time series for each dimension). Each times series is then plotted against all others and the dynamic patterns of the unfolded gait time series' are examined. For example, a one-dimensional knee angle time series can be unfolded into a two-dimensional or three-dimensional time series as shown below (Figure 2).

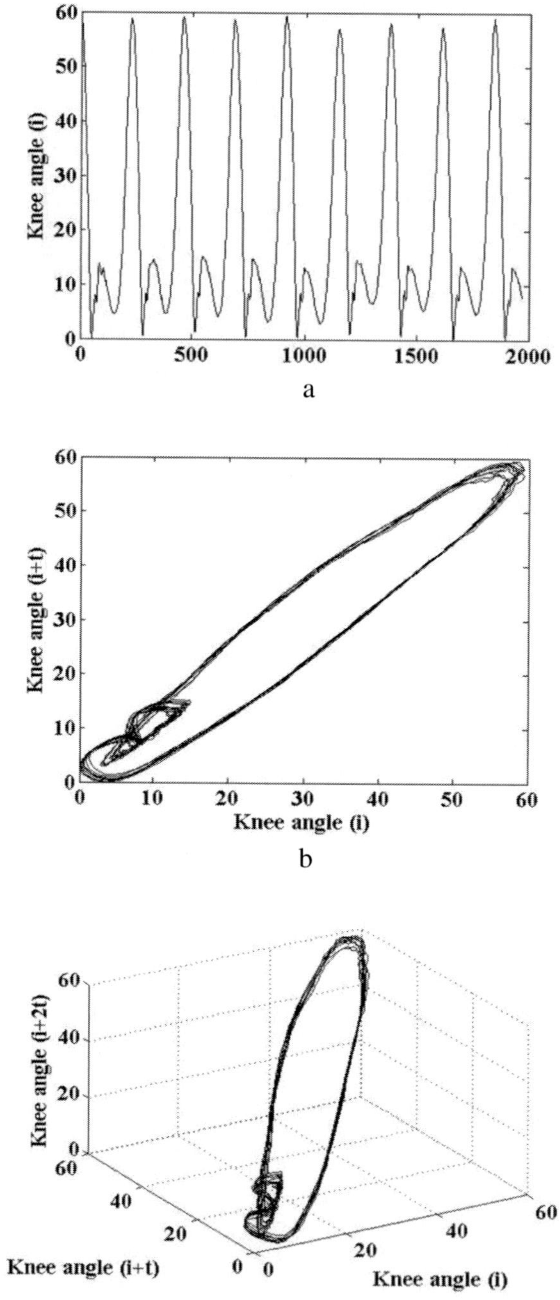

Figure 2. Knee angle time series in one dimension (a), two dimensions (b), and three dimensions (c). A time delay of 5 data points was used for reconstruction of the multidimensional state space in two and three dimensions.

Although we are limited to the visual exploration of only a one-, two-, or three-dimensional state space, the unfolding process can continue into *n*-dimensions, and this allows for a mathematical exploration of high-dimensional dynamic processes. For continuous knee angle data during gait, a d_E=5 is commonly used (Buzzi and Ulrich, 2004;

Dingwell, et al., 2000). However, each time series should be independently evaluated to determine the proper d_E.

A popular computational method to examine the dynamic patterns of the state space is the Lyapunov exponent, which quantifies the rate of divergence in neighboring trajectories within the multidimensional state-space. Larger positive numbers designate greater instability, as this indicates rapid divergence of the system over time and that the system is susceptible to local perturbations. The maximum Lyapunov exponent (λ_1) can only be reliably calculated from extremely long data sets that are not practical in experimental or clinical settings. Thus, the maximum finite-time Lyapunov exponent (λ^*) was developed (Rosenstein, Collins, and DeLuca, 1993) and further divided into λ^*_S and λ^*_L to examine the dynamic characteristics over short and long time scales, respectively (Dingwell and Kang, 2007). λ^*_S and λ^*_L are determined by first calculating the relative distance of the nearest neighboring points in the multidimensional state-space and then a determination of the line of best fit at different time scales is made with the following equation:

$$y(i) = \frac{1}{\Delta t}\langle \ln[d_j(i)] \rangle, \qquad (5)$$

where $d_j(i)$ is the Euclidean distance between the jth pair of nearest neighbors after i discrete time steps ($i\Delta t$ seconds) and $\langle \cdot \rangle$ the average over all values of j (Rosenstein, et al., 1993). λ^*_S and λ^*_L are typically calculated over 1 stride cycle and 4-10 stride cycles, respectively, and provide a measure of local dynamic stability at different time scales (Figure 3). If λ^*_S and λ^*_L tend toward 0, then no divergence between the state-space trajectories is observed and the system is considered to be stable. Thus, larger values indicate instability.

The local dynamic stability of gait behavior is consistent with the dynamic disease framework in that the dynamics of behavior are altered when constraints are altered. For example, constraints on the individual such as normal aging, ACL rupture, knee osteoarthritis, and Down's syndrome result in lower local dynamic stability (Buzzi, Stergiou, Kurz, Hageman, and Heidel, 2003; Buzzi and Ulrich, 2004; Moraiti, Stergiou, Ristanis, and Georgoulis, 2007; Yakhdani et al., 2012). Task constraints, such as changes in walking speed or different terrain, also affect the local dynamic stability of gait dynamics (Bruijn, Dieën, Meijer, and Beek, 2009; Dingwell and Cusumano, 2000; Dingwell and Kang, 2007). These studies provide convergent evidence that functional mobility emerges as a function of the interaction of constraints on gait dynamics.

Dynamic Recurrence

All previous sections have identified unique ways to quantify the repeating patterns in gait dynamics. Another way to conceptualize repeating patterns is to determine when the dynamics of a multidimensional state space pass through the same geometrical space, with each instance of this termed a recurrence.

Recurrent dynamics are a hallmark of a healthy, adaptive physiological system. As previously discussed, a functional system exhibits both repeatability and adaptive variation. Thus, a system with too much or too little recurrence may be maladaptive at a functional level.

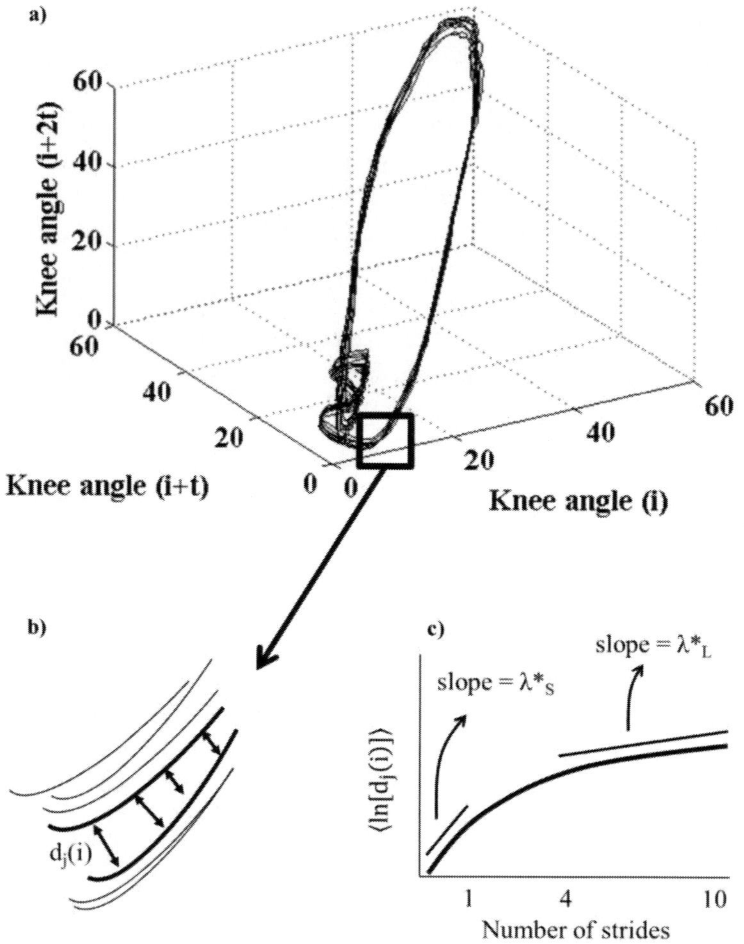

Figure 3. Representation of the calculation of Lyapunov exponent. The three dimensional knee angle time series (a) is zoomed in on to show that neighboring trajectories are converging over a short window of time (b). The distance of two neighboring trajectories at each time sample is calculated ($d_j(i)$) and plotted over a 10 step interval. The slope of the log of the distance over one step (λ^*_S) and between 4-10 steps (λ^*_L) is a measure of the local dynamic stability over a short and long time scale, respectively (c).

A popular way to index recurrence in physiological systems is through recurrence quantification analysis (RQA) (Webber and Zbilut, 1994); an analysis based on the conceptualization of recurrence plots (Eckmann, Kamphorst, and Ruelle, 1987). Similar to the dynamic stability classification described in the previous section, RQA is predicated on the theorem derived by Takens (1981)—the nature of a multidimensional system can be understood through comparison of time-delayed copies of the original data so that the dynamics of the system can be properly unfolded. Accordingly, time-delayed copies prescribed in Equation 4 are generated and plotted against the original time series.

The embedding dimension and time delay for this process is identical to what was described in the previous section of this chapter. Therefore, the initial steps for the computation of dynamic stability and recurrence are the same.

Once the multidimensional state-space is created, RQA can be employed to examine several dynamic properties in the unfolded gait time series. This is an advantage RQA has over other techniques, as most only quantify one property of the gait dynamics. RQA was originally developed so that four properties of the dynamics could be indexed: (1) percent recurrence (%REC), (2) percent determinism (%DET), (3) maxline (MAXL), and (4) entropy (ENT)[2]. Since its inception, many more dynamic properties of the recurrence plot have been formalized (Marwan, 2011; Marwan, Romano, Thiel, and Kurths, 2007; Webber and Zbilut, 2005; Zbilut and Webber, 2008).

The first step in RQA is to define the number of recurrent points (i.e., neighbors) in the multidimensional state-space. This is done by defining a radius (r_{rqa})[3], which acts as a sensitivity parameter that determines the existence of a recurrent point in Euclidean space. If r_{rqa} is very large, then all points in the multidimensional state-space would recur with each other and no discernable dynamics of the system could be quantified. Alternatively, if r_{rqa} is very small, then no recurrent values can be defined.

Therefore, an appropriately sized r_{rqa} is essential to extract potentially meaningful information from the recurrence plot. As the values of an appropriate r_{rqa} will be dependent on the system of study, it is suggested that the size of r_{rqa} should be selected so that %REC remains appropriately low, and a %REC of approximately 5% is a good baseline (Pellechia and Shockley, 2005).

It should be noted that there is no standard in the literature about whether r_{rqa} should remain consistent between subjects, tasks, or conditions (which would allow %REC to fluctuate), or if a different r should be selected for every subject, task, or condition (this would render %REC stable). This means that great care should be taken in the selection of the r_{rqa} parameter, as each subsequent output measure (i.e., %DET, ENT, and MAXL) is conditional on the identification of recurrent points within the observed system. Once an appropriate r_{rqa} is selected, the number of recurrent points in the system (%REC) can be calculated. Qualitatively, this can be expressed in a recurrence plot as seen in the Figure 4.

The darkened points on the plot represent each instance when two points are neighbors in state space. A distinctive feature of the recurrence plot is that a diagonal line (termed the line of identity) separates the plot into two equal triangles. The bottom and top triangle of the plot are symmetrical (similar to a correlation table), so only half of the plot is used in the subsequent analysis, excluding the line of identity itself.

Quantitatively, the number of recurrent points divided by the total number of possible recurrent points equates to %REC (i.e., if the entire plot was black, %REC would be 100%). Therefore, %REC is kept low so that the embedded dynamical structure of the observed system can be extracted from the recurrence plot.

[2] Entropy from RQA is a different construct than what was previously described in the ApEn and SampEn analyses. RQA entropy is based on Shannon entropy, while the other two are based on K-S entropy. For a description of the different classes of entropy see Williams (1997) and Rhea et al. (2011).

[3] The radius parameter for RQA is typically denoted as r. Since this nomenclature is the same as the level of tolerance (r) outlined in the entropy section, we've elected to denote the RQA radius parameter as r_{rqa} to prevent confusion. While r and r_{rqa} are parameters that act as a sensitivity metric to guide the selection of repeating patterns, their actual meaning within the two analyses, along with how they are selected is very different.

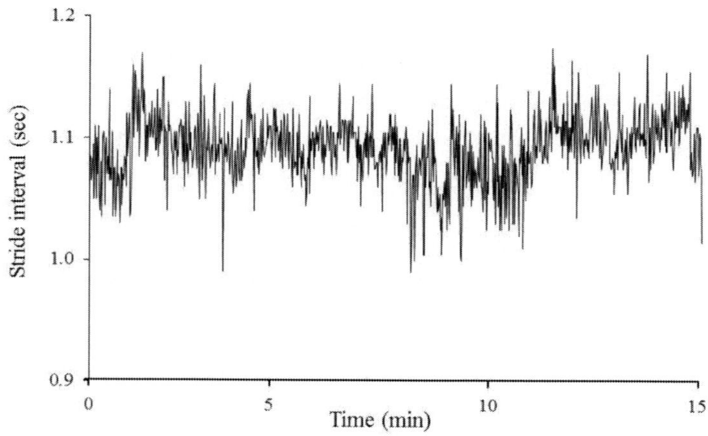

a

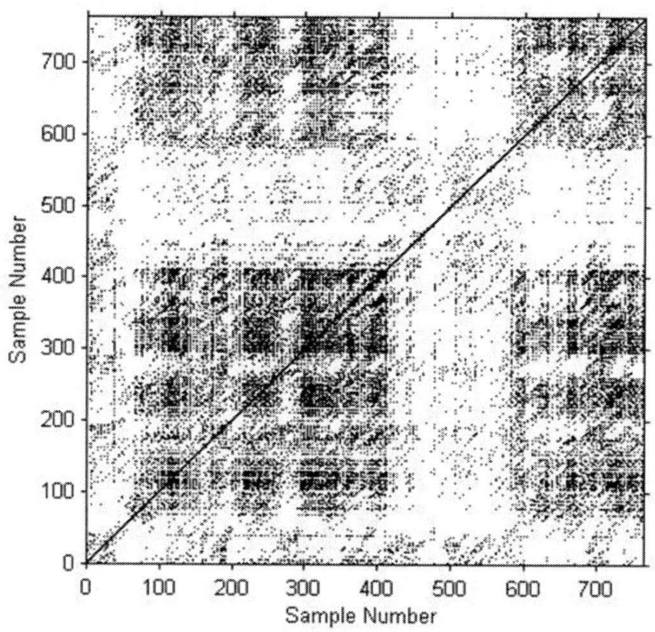

b

Figure 4. Recurrence plot. A stride interval time series (a) was transformed into a recurrence plot using an embedding dimension of 5, time delay of 5, radius of 45 and line minimum of 2. The resulting RQA output values were: %REC=4.7%, %DET=15.7%, ENT=0.6, and MAXL=9.

While a number of output variables can be computed from the recurrence plot, we choose to focus on the three variables that are most often reported. %DET is a measurement of regularity within the recurrence plot and is calculated by dividing the number of points that fall on a diagonal line (typically defined by 2 or more points in a row) by the total number of recurrent points.

ENT indexes the dynamic patterns of the diagonal lines by taking the negative sum of the product of the probability of line lengths multiplied by $\log_2$ of the probability of line lengths.

Lastly, MAXL is simply the number of data points in the longest diagonal line independent of the line of identity, and has been reported to be inversely related to the Lyapunov exponent (Trulla, Giuliani, Zbilut, and Webber, 1996), although the strength of that relation has recently come into question (Marwan, 2011). In the context of human motor behavior, RQA has mostly been employed in postural control research (Balasubramaniam, Riley, and Turvey, 2000; Hasson et al., 2008; Mazaheri, Negahban, Salavati, Sanjari, and Parnianpour, 2010; Negahban et al., 2010; Riley, Balasubramaniam, and Turvey, 1999; Riley and Clark, 2003; Schmit, Regis, and Riley, 2005; Schmit et al., 2006; Seigle, Ramdani, and Bernard, 2009). However, RQA has recently been utilized to examine the dynamic patterns of gait (Labini, Meli, Ivanenko, and Tufarelli, 2012; Rhea, et al., under review). All of these results support the dynamical disease framework in that changes to the constraints imposed on the system lead to a change in performance dynamics. While RQA is attractive because it allows for a greater description of the system through the characterization of several components of the gait dynamics, much work needs to be done to discover how RQA variables relate to functional mobility.

Multifractal Computational Methods

All of the previously described computational methods share one common characteristic—a metric is produced that indexes patterns within the signal. This implies that the same patterns are observed throughout the various time scales, a characteristic that is described as fractal (Mandelbrot, 1982). The original ideas of Mandelbrot were delineated as monofractal patterns after recent methodological advances have shown that the dynamics of many systems interact at time scales of different lengths—a characteristic that is described as multifractal (Glass, 2001; Goldberger et al., 2002; Ivanov, et al., 1999; Stanley, et al., 1999). Thus, monofractal analyses may fall short due to their inability to index the richness of the dynamic patterns across a variety of time scales. To this end, some monofractal analyses have been extended to index multifractality. For example, DFA has been extended to multifractal DFA (MF-DFA) (Kantelhardt et al., 2002). Similarly, a multi-fractal version of the entropy method has been developed and is called multiscale entropy (MSE) (Costa, Goldberger, and Peng, 2002). MSE has been applied to gait data and further confirmed that the patterns observed in gait dynamics arise at different time scales and are influenced by constraints imposed on the system (Costa, Peng, Goldberger, and Hausdorff, 2003). Future research should focus on employing multifractal methods to more accurately index gait dynamics across time scales, which may lead to enhanced diagnostic and rehabilitative metrics.

Dynamical Disease Framework for Clinical Rehabilitation

The majority of this chapter described an extension of the dynamical disease framework to understand the variable nature of gait, how gait dynamics can be quantified, and the implications of these measures for functional behavior.

Much of the early work in this area was descriptive—a necessary first step to understand how constraints such as a neurological insult or changing task demands influence gait dynamics. Each of the computational methods discussed quantifies a unique component of the gait dynamics. A challenge for future research is to continue to explore the relation between computational methods used to index patterns in gait variability, and to integrate these measures to provide a broader picture of functional mobility. From a clinical perspective, it would be extremely valuable to compute a group of variables for a defined set of tasks that can accurately predict functional mobility. This would be the next logical step for the utility of dynamical disease in clinical practice because it would allow for an objective, quantifiable way to access and track patient progress. Furthermore, this framework could be used to enhance current clinical interventions. For example, a patient may exhibit a metric of gait variability that deviates from normal (e.g., weak long-range correlations), and this might predict a decrement in functional mobility. The underlying component(s) known to influence such a change could then be targeted by their rehabilitation. This has motivated researchers to explore methods that alter the gait dynamics so that functional mobility can re-emerge in clinical populations (Hove, Suzuki, Uchitomi, Orimo, and Miyake, 2012; Rhea, Kiefer, Warren, D'Andrea, and Aaron, 2011). Advances in technology to deliver a variety of stimuli aimed at shifting patients toward normal gait dynamics, along with increased portability of measurement devices necessary to index these changes, provides much promise for the progression of a dynamical disease framework toward a more prominent role for clinical rehabilitation.

A necessary area of advancement prior to the implementing a dynamic disease framework for clinical rehabilitation is a stronger understanding of how constraints influence a system's dynamics. The patient's motor behavior is a function of the constraints imposed on actor-task-environment system (Newell, 1986). A component-dominant approach supposes that each component of the system adds a unique characteristic to the behavioral dynamics. Alternatively, an interaction-dominant approach postulates that the behavioral dynamics arise from the nonlinear interaction among the system's components (Diniz, et al., 2011; Van Orden, Holden, and Turvey, 2003). From a rehabilitation perspective, the interaction-dominant approach implies altering constraints on one part of the system (i.e., task difficulty) may change the behavioral dynamics in a nonlinear manner. This provides a challenge for researchers interested in utilizing a dynamical disease framework for rehabilitation, as it will be important to understand how the dynamics of a rehabilitation program can be manipulated in a way that is beneficial for the patient.

One roadblock to the adoption of this approach into the mainstream of rehabilitation practice is that it requires relatively large data sets compared to traditional clinical assessments. Moreover, an advanced understanding of mathematics is necessary to perform and understand the analyses. The nature of the dynamical disease framework will always require a large number of data points in order for behavioral patterns to be well understood. However, the number of data points needed to accurately and reliability quantify gait patterns remains an open empirical question.

Clinical rehabilitation can benefit from research that focuses on the development of the analyses, and the subsequent metrics that can be consistently obtained in short data sets. As the field progresses, a more fundamental understanding of the underlying characteristics in gait dynamics will be obtained, hopefully in parallel with an understanding of how such characteristics relate to functional mobility.

In due time, standards of how to process dynamical data, as well as how to interpret the metrics of each computational method will need to be created. Once these practices are standardized, less importance will be placed on understanding the mathematical rigor required to perform a gait dynamics analysis. Just as most people are able to utilize a car even without a fine-grained understanding of how a car engine works, the dynamical disease framework may someday be simplified into a simple “red light / green light” test in the clinic. A physical therapist would follow the standardized guidelines for the task protocol and the computer program would analyze and interpret the gait behavior to indicate if the patient exhibited adaptive (green light) or maladaptive (red light) gait behavior.

Of course, a test of this nature would be used in conjunction with other standard clinical tests to gain a holistic view of the patients functional gait ability. This approach would allow for the evolution of the dynamical disease framework into an everyday clinical setting and provide access to clinicians who do not have formal training in dynamic analyses, but this may come with the unintended consequence of the misinterpretation of a patient’s behavior. Because of this, it is imperative that researchers move toward a unified language and interpretation of dynamic analyses as the field develops.

Conclusion

It is clear that gait variability should no longer be discarded as random fluctuations in motor control. Much has been learned about the dynamic nature of gait variability over the past two decades in a variety of contexts. The dynamic disease framework allows for a holistic understanding of how constraints influence the gait dynamics, which in turn influences the ability of the system to respond to internal and external demands. This chapter covered four classes of computational methods commonly used to analyze gait dynamics, including long-range correlations, entropy, local dynamic stability, and dynamic recurrence. Each class allows for a different property of the gait dynamics to be indexed.

However, it is unclear which property or properties indicate the presence of functional mobility. Future research should continue to pursue the relation between dynamic patterns in gait and functional mobility to aid researchers and clinicians in deciding which metric(s) provide the most relevant information for evidence-based practice. The dynamical disease framework can not only be used to assess a person’s functional ability, but also to enhance clinical rehabilitation of gait disorders. This evolution is dependent upon first describing which measures of gait dynamics are most sensitive to motor control dysfunction so that clinical experts can prescribe protocols to specifically target the component of the locomotor system tied to the dysfunction. The challenge for current and future researchers is to develop a unified approach of best practices for a dynamics framework to understand and treat human mobility dysfunction.

References

Adams, J. A. (1971). A closed-loop theory of motor learning. *Journal of Motor Behavior,* 3, 111-150.

Arif, M., Ohtaki, Y., Nagatomi, R., and Inooka, H. (2004). Estimation of the effect of cadence on gait stability in young andelderly people using approximate entropy technique. *Measurement Science Review,* 4(2), 29-40.

Arif, M., Othaki, Y., Nagatomi, R., Ishihara, T., and Inooka, H. (2002). Analysis of the effect of fatigue on walking stability. *Proceedings of International Symposium on Micromechatronics and Human Science IEEE*, 253-258.

Balasubramaniam, R., Riley, M. A. and Turvey, M. T. (2000). Specificity of postural sway to the demands of a precision task. *Gait and Posture,* 11(1), 12-24.

Belair, J., Glass, L., Heiden, U., and Milton, J. (1995). Dynamical disease: Identification, temporal aspects and treatment strategies of human illness. *Chaos,* 5(1).

Berg, K., Wood-Dauphinee, S., Williams, J. I., and Gayton, D. (1989). Measuring balance in the elderly: preliminary development of an instrument. *Physiotherapy Canada,* 41, 304-311.

Bravi, A., Longtin, A. and Seely, A. J. E. (2011). Review and classification of variabilitiy analysis techniques with clinical applications. *BioMedical Engineering OnLine,* 10, 90.

Bruijn, S. M., Bregman, D. J. J., Meijer, O. G., Beek, P. J., and Dieen, J. H. v. (2012). Maximum Lyapunov exponents as predictors of global gait stability: A modeling approach. *Medical Engineering and Physics,* 34(4), 428-436.

Bruijn, S. M., Dieën, J. H. v., Meijer, O. G., and Beek, P. J. (2009). Statistical precision and sensitivity of measures of dynamic gait stability. *Journal of Neuroscience Methods,* 178 (2), 327-333.

Buzzi, U. H., Stergiou, N., Kurz, M. J., Hageman, P. A., and Heidel, J. (2003). Nonlinear dynamics indicates aging affects variability during gait. *Clinical Biomechanics,* 18, 435-443.

Buzzi, U. H. and Ulrich, B. D. (2004). Dynamic stability of gait cycles as a function of speed and system constraints. *Motor Control,* 8(3), 241-254.

Cavanaugh, J. T., Guskiewicz, K. M., Giuliani, C., Marshall, S., Mercer, V. S., and Stergiou, N. (2006). Recovery of postural control after cerebral concussion: New insights using approximate entropy. *Journal of Athletic Training,* 41(3), 305-313.

Costa, M., Goldberger, A. L. and Peng, C. K. (2002). Multiscale entropy analysis of complex physiologic time series. *Physical Review Letters,* 89(6), 068102.

Costa, M., Peng, C.-K., Goldberger, A. L., and Hausdorff, J. M. (2003). Multiscale entropy analysis of human gait dynamics. *Physica A,* 330, 53-60.

Diedrich, F. J. and Warren, W. H. (1995). Why change gaits? Dynamics of the walk-run transition. *Journal of Experimental Psychology: Human Perception and Performance,* 21 (1), 183-202.

Dingwell, J. B. and Cusumano, J. P. (2000). Nonlinear time series analysis of normal and pathological human walking. *Chaos,* 10(4), 848-863.

Dingwell, J. B., Cusumano, J. P., Sternad, D., and Cavanagh, P. R. (2000). Slower speeds in patients with diabetic neuropathy lead to improved local dynamic stability of continuous overground walking. *Journal of Biomechanics,* 33(10), 1269-1277.

Dingwell, J. B. and Kang, H. G. (2007). Differences between local and orbital dynamic stability during human walking. *Journal of Biomechanical Engineering,* 129(4), 586-593.

Dingwell, J. B. and Marin, L. C. (2006). Kinematic variability and local dynamic stability of upper body motions when walking at different speeds. *Journal of Biomechanics,* 39(3), 444-452.

Dingwell, J. B., Robb, R. T., Troy, K. L., and Grabiner, M. D. (2008). Effects of an attention demanding task on dynamic stability during treadmill walking. *Journal of Neuro Engineering and Rehabilitation,* 5(12).

Diniz, A., Wijnants, M. L., Torre, K., Barreiros, J., Crato, N., Bosman, A. M. T., ... Delignieres, D. (2011). Contemporary theories of 1/f noise in motor control. *Human Movement Science,* 30, 889-905.

Donker, S. F., Ledebt, A., Roerdink, M., Savelsbergh, G. J. P., and Beek, P. J. (2008). Children with cerebral palsy exhibit greater and more regular postural sway than typically developing children. *Experimental Brain Research,* 184(3), 363-370.

Duarte, M. and Zatsiorsky, V. M. (2001). Long-range correlations in human standing. *Physics Letters A,* 283, 124-128.

Eckmann, J., Kamphorst, S. and Ruelle, D. (1987). Recurrence plots of dynamical systems. *Europhysics Letters,* 5(9), 973-977.

Fraser, A. M. and Swinney, H. L. (1986). Independent cordinates for strange attractors from mutual information. *Physical Review A,* 33, 1134-1140.

Gates, D. H. and Dingwell, J. B. (2007). Peripheral neuropathy does not alter the fractal dynamics of stride intervals of gait. *Journal of Applied Physiology,* 102(3), 965-971.

Georgoulis, A. D., Moraiti, C., Ristanis, S., and Stergiou, N. (2006). A novel approach to measure variability in the anterior cruciate ligament deficient knee during walking: the use of the approximate entropy in orthopaedics. *Journal of Clinical Monitoring and Computing,* 20(1), 11-18.

Glass, L. (2001). Synchronization and rhythmic processes in physiology. *Nature,* 410(6825), 277-284.

Glass, L. and Mackey, M. C. (1979). Pathological conditions resulting from instabilities in physiological control systems. *Annals of the New York Academy of Sciences,* 316, 214-235.

Goldberger, A. L. (1991). Is the normal heartbeat chaotic or hemostatic. *News in Physiological Sciences,* 6, 87-91.

Goldberger, A. L., Amaral, L. A. N., Hausdorff, J. M., Ivanov, P. C., Peng, C.-K., and Stanley, H. E. (2002). Fractal dynamics in physiology: Alterations with disease and aging. *Proceedings from the National Academy of Sciences of The US* 99(Suppl. 1), 2466-2476.

Goldberger, A. L., Rigney, D. R. and West, B. J. (1990). Chaos and fractals in human physiology. *Scientific American,* 262, 42-49.

Goldberger, A. L. and West, B. J. (1987). Applications of nonlinear dynamics to clinical cardiology. *Annals of the New York Academy of Sciences,* 504, 195-213.

Govindan, R. B., Wilsona, J. D., Eswaran, H., Lowery, C. L., and Preißl, H. (2007). Revisiting sample entropy analysis. *Physica A: Statistical Mechanics and its Applications,* 376, 158-164.

Harbourne, R. T. and Stergiou, N. (2009). Movement Variability and the Use of Nonlinear Tools: Principles to Guide Physical Therapist Practice. *Physical Therapy,* 89(3), 267-282.

Hasson, C. J., Emmerik, R. E. A. V., Caldwell, G. E., Haddad, J. M., Gagnon, J. L., and Hamill, J. (2008). Influence of embedding parameters and noise in center of pressure recurrence quantification analysis. *Gait and Posture* 27(3), 416-422.

Hausdorff, J. M. (2007). Gait dynamics, fractals and falls: Finding meaning in the stride-to-stride fluctuations of human walking. *Human Movement Science,* 26(4), 555-589.

Hausdorff, J. M., Cudkowicz, M. E., Firtion, R., Wei, J. Y., and Goldberger, A. L. (1998). Gait variability and basal ganglia disorders: Stride-to-stride variations of gait cycle timing in Parkinson's disease and Huntington's disease. *Movement Disorders,* 13(3), 428-437.

Hausdorff, J. M., Lertratanakul, A., Cudkowicz, M. E., Peterson, A. L., Kaliton, D., and Goldberger, A. L. (2000). Dynamic markers of altered gait rhythm in amyotrophic lateral sclerosis. *Journal of Applied Physiology,* 88(6), 2045-2053.

Hausdorff, J. M., Mitchell, S. L., Firtion, R., Peng, C. K., Cudkowicz, M. E., Wei, J. Y., and Goldberger, A. L. (1997). Altered fractal dynamics of gait: reduced stride-interval correlations with aging and Huntington's disease. *Journal of Applied Physiology,* 82(1), 262-269.

Hausdorff, J. M., Peng, C.-K., Ladin, Z., Wei, J. Y., and Goldberger, A. L. (1995). Is walking a random walk? Evidence for long-range correlations in stride interval of human gait. *Journal of Applied Physiology,* 78(1), 349-358.

Hausdorff, J. M., Purdon, P. L., Peng, C. K., Ladin, Z., Wei, J. Y., and Goldberger, A. L. (1996). Fractal dynamics of human gait: stability of long-range correlations in stride interval fluctuations. *Journal of Applied Physiology,* 80(5), 1448-1457.

Hausdorff, J. M., Zemany, L., Peng, C.-K., and Goldberger, A. L. (1999). Maturation of gait dynamics: stride-to-stride variability and its temporal organization in children. *Journal of Applied Physiology,* 86(3), 1040-1047.

Hove, M. J., Suzuki, K., Uchitomi, H., Orimo, S., and Miyake, Y. (2012). Interactive rhythmic auditory stimulation reinstates natural 1/*f* timing in gait of Parkinson's patients. *PLoS ONE,* 7(3), e32600.

Hurd, W. J. and Snyder-Mackler, L. (2007). Knee instability after acute ACL rupture affects movement patterns during the mid-stance phase of gait. *Journal of Orthopaedic Research,* 25(10), 1369-1377.

Ivanov, P. C., Amaral, L. A. N., Goldberger, A. L., Havlin, S., Rosenblum, M. G., Struzik, Z. R., and Stanley, H. E. (1999). Multifractality in human heartbeat dynamics. *Nature,* 399, 461-465.

Jordan, K., Challis, J. H., Cusumano, J. P., and Newell, K. M. (2009). Stability and the time-dependent structure of gait variability in walking and running. *Human Movement Science,* 28(1), 113-128.

Jordan, K., Challis, J. H. and Newell, K. M. (2006). Long range correlations in the stride interval of running. *Gait and Posture* 24(1), 120-125.

Jordan, K., Challis, J. H. and Newell, K. M. (2007a). Speed influences on the scaling behavior of gait cycle fluctuations during treadmill running. *Human Movement Science,* 26(1), 87-102.

Jordan, K., Challis, J. H. and Newell, K. M. (2007b). Walking speed influences on gait cycle variability. *Gait and Posture,* 26(1), 128-134.

Kantelhardt, J. W., Zschiegner, S. A., Koscielny-Bunde, E., Havlin, S., Bunde, A., and Stanley, H. E. (2002). Multifractal detrended fluctutation analysis of nonstationary time series. *Physica A,* 316, 87-114.

Kaplan, D. T., Furman, M. I., Pincus, S. M., Ryan, S. M., Lipsitz, L. A., and Goldberger, A. L. (1991). Aging and the complexity of cardiovascular dynamics. *Biophysical Journal,* 59, 945-949.

Katsavelis, D., Mukherjee, M., Decker, L., and Stergiou, N. (2010a). The effect of virtual reality on gait variability. *Nonlinear Dynamics, Psychology, and Life Sciences,* 14(3), 239-256.

Katsavelis, D., Mukherjee, M., Decker, L., and Stergiou, N. (2010b). Variability of lower extremity joint kinematics during backward walking in a virtual environment. *Nonlinear Dynamics, Psychology, and Life Sciences.,* 14(2), 165-178.

Kennel, M. B., Brown, R. and Abarbanel, H. D. (1992). Determining embedding dimension for phase-space reconstruction using a geometrical construction. *Physical Review A,* 45 (6), 3403-3411.

Kiefer, A. W., Riley, M. A., Shockley, K., Villard, S., and Van Orden, G. C. (2009). Walking changes the dynamics of cognitive estimates of time intervals. *Journal of Experimental Psychology: Human Perception and Performance,* 35(5), 1532-1541.

Kiefer, A. W., Wallot, S., Gresham, L. J., Kloos, H., Riley, M. A., Shockley, K., and Van Orden, G. C. (under review). Development of coordination in time estimation. *Developmental Psychology.*

King, G. P., Jones, R. and Broomhead, D. S. (1987). Phase portraits from a time series: a singular system approach. *Nuclear Physics B,* 2, 379-390.

Kloos, H. and Van Orden, G. C. (2010). Voluntary behavior in cognitive and motor tasks. *Mind and Matter,* 8(1), 19-43.

Labini, F. S., Meli, A., Ivanenko, Y. P., and Tufarelli, D. (2012). Recurrence quantification analysis of gait in normal and hypovestibular subjects. *Gait and Posture,* 35, 48-55.

Lamoth, C. J. C., Ainsworth, E., Polomski, W., and Haudijk, H. (2010). Variability and stability analysis of walking of transfemoral amputees. *Medical Engineering and Physics,* 32, 1009-1014.

Lindstrom, M., Fellander-Tsai, L., Wredmark, T., and Henriksson, M. (2010). Adaptations of gait and muscle activiation in chronic ACL deficiency. *Knee Surgery, Sports Traumatology, Arthroscopy,* 18, 106-114.

Lipsitz, L. A. (2002). Dynamics of stability: The physiologic basis of functional health and frailty. *Journal of Gerontology: Biological Sciences,* 57A(3), B115-B125.

Lipsitz, L. A. and Goldberger, A. L. (1992). Loss of 'complexity' and aging: potential applications of fractals and chaos theory to senescence. *JAMA: The Journal of the American Medicine Association,* 267(13), 1806-1809.

Mackey, M. C. and Glass, L. (1977). Oscillation and chaos in physiological control systems. *Science,* 197, 287-289.

Mackey, M. C. and Milton, J. G. (1987). Dynamical diseases. *Annals of the New York Academy of Sciences,* 504, 16-32.

Mandelbrot, B. B. (1982). *The Fractal Geometry of Nature*. New York: Freeman.

Marwan, N. (2011). How to avoid potential pitfalls in recurrence plot based data analysis. *International Journal of Bifurcation and Chaos,* 21, 1003.

Marwan, N., Romano, M. C., Thiel, M., and Kurths, J. (2007). Recurrence plots for the analysis of complex systems. *Physics Reports* 438(5-6), 237-329.

Mazaheri, M., Negahban, H., Salavati, M., Sanjari, M. A., and Parnianpour, M. (2010). Reliability of recurrence quantification analysis measures of the center of pressure during standing in individuals with musculoskeletal disorders. *Medical Engineering and Physics* 32(7), 808-812.

Moraiti, C. O., Stergiou, N., Ristanis, S., and Georgoulis, A. D. (2007). ACL deficiency affects stride-to-stride variability as measured using nonlinear methodology. *Knee Surgery, Sports Traumatology, Arthroscopy,* 15(12), 1406-1413.

Moraiti, C. O., Stergiou, N., Ristanis, S., Vasiliadis, H. S., Patras, K., Lee, C., and Georgoulis, A. D. (2009). The effect of anterior cruciate ligament reconstruction on stride-to-stride variability. *Arthroscopy,* 25(7), 742-749.

Negahban, H., Salavati, M., Mazaheri, M., Sanjari, M. A., Hadian, M. R., and Parnianpour, M. (2010). Non-linear dynamical features of center of pressure extracted by recurrence quantification analysis in people with unilateral anterior cruciate ligament injury. *Gait and Posture,* 31(4), 450-455.

Newell, K. M. (1986). Constraints on the development of coordination. In: M. G. Wade and H. T. A. Whiting (Eds.), *Motor skill acquisition in children: Aspects of coordination and control* (pp. 341-360). Amsterdam: Martinies NIJHOS.

Olmo, M. F. d. and Cudeiro, J. (2005). Temporal variability of gait in Parkinson disease: Effects of a rehabilitation programme based on rhythmic sound cues. *Parkinsonism and Related Disorders,* 11, 25-33.

Osternig, L. R., Caster, B. L. and James, C. R. (1995). Contralateral hamstring (biceps femoris) coactivation patterns and anterior cruciate ligament dysfunction. *Medicine and Science in Sports and Exercise,* 27(6), 805-808.

Pellechia, G. L. and Shockley, K. (2005). Application of recurrence quantification analysis: Influence of cognitive activity on postural fluctuations. In: M. A. Riley and G. C. Van Orden (Eds.), *Tutorials in contemporary nonlinear methods for the behavioral sciences* (pp. 95-141): NSF.

Peng, C. K., Buldyrev, S. V., Goldberger, A. L., Havlin, S., Sciortino, F., Simons, M., and Stanley, H. E. (1992). Long-range correlations in nucleotide sequences. *Nature,* 356, 168-170.

Peng, C. K., Buldyrev, S. V., Havlin, S., Simons, M., Stanley, H. E., and Goldberger, A. L. (1994). Mosaic organization of DNA necleotides. *Physical Review E,* 49(2), 1685–1689.

Peng, C. K., Havlin, S., Stanley, H. E., and Goldberger, A. L. (1995). Quantification of scaling exponents and crossover phenomena in nonstationary heartbeat time series. *Chaos,* 5(1), 82-87.

Peng, C. K., Mietus, J., Hausdorff, J. M., Havlin, S., Stanley, H. E., and Goldberger, A. L. (1993). Long-range anticorrelations and non-Gaussian behavior of the heartbeat. *Physical Review Letters,* 79(9), 1343-1346.

Perlmutter, S., Lin, F. and Makhsous, M. (2010). Quantitative analysis of static sitting posture in chronic stroke. *Gait and Posture,* 32(1), 53-56.

Pincus, S. M. (1991). Approximate entropy as a measure of system complexity. *Proc. Natl. Acad. Sci. US,* 88(6), 2297–2301.

Ramdani, S., Bouchara, F. and Lagarde, J. (2009). Influence of noise on the sample entropy algorithm. *Chaos,* 19(1), 013123.

Reeves, N. P., Narendra, K. S. and Cholewicki, J. (2007). Spine stability: the six blind men and the elephant. *Clinical Biomechanics,* 22(3), 266-274.

Rhea, C. K., Kiefer, A. W., D'Andrea, S. E., Warren, W. H., and Aaron, R. K. (2010). *Approximate entropy of stride-to stride intervals following ACL injury.* Paper presented at the American Society for Biomechanics, Providence, RI.

Rhea, C. K., Kiefer, A. W., Gerin-Lajoie, M., Warren, W. H., and Aaron, R. K. (under review). Constraining knee motion changes the dynamic structure of the direction of travel in a functional mobility task.

Rhea, C. K., Kiefer, A. W., Warren, W. H., D'Andrea, S. E., and Aaron, R. K. (2011). *Synchronizing to a "noisy" metronome induces corresponding shifts in fractal gait dynamics.* Paper presented at the North American Society for Psychology in Sport and Physical Activity, Burlington, VT.

Rhea, C. K., Silver, T. A., Hong, S. L., Ryu, J. H., Studenka, B. E., Hughes, C. M. L., and Haddad, J. M. (2011). Noise and complexity in human postural control: Interpreting the different estimations of entropy. *PLoS ONE,* 6(3), e17696.

Richman, J. S. and Moorman, J. R. (2000). Physiological time-series analysis using approximate entropy and sample entropy. *Am. J. Physiol. Heart Circ. Physiol.,* 278(6), H2039-H2049.

Riley, M. A., Balasubramaniam, R. and Turvey, M. T. (1999). Recurrence quantification analysis of postural fluctuations. *Gait and Posture* 9(1), 65-78.

Riley, M. A. and Clark, S. (2003). Recurrence analysis of human postural sway during the sensory organization test. *Neuroscience Letters,* 342(1-2), 45-48.

Roerdink, M., Haart, M. D., Daffertshofer, A., Donker, S. F., Geurts, A. C. H., and Beek, P. J. (2006). Dynamical structure of center-of-pressure trajectroies in patients recovering from stroke. *Experimental Brain Research,* 174, 256-269.

Rosenstein, M. T., Collins, J. and Luca, C. D. (1994). Reconstruction expansion as a geometry-based framework for choosing proper delay times. *Physica D,* 73, 82-98.

Rosenstein, M. T., Collins, J. J. and DeLuca, C. J. (1993). A practical method for calculating largest Lyapunov exponents from small data sets. *Physica D,* 65, 117-134.

Rudolph, K., Axe, M., Buchanan, T., Scholz, J., and Snyder-Mackler, L. (2001). Dynamic stability in the anterior cruciate ligament deficient knee. *Knee Surgery, Sports Traumatology, Arthroscopy,* 9(2), 62-71.

Scafetta, N., Marchi, D. and West, B. J. (2009). Understanding the complexity of human gait dynamics. *Chaos,* 19, 026108.

Schmidt, R. A. (1975). A schema theory of discrete motor skill learning. *Psychological Review,* 82(4), 225-260.

Schmit, J. M., Regis, D. I. and Riley, M. A. (2005). Dynamic patterns of postural sway in ballet dancers and track athletes. *Experimental Brain Research,* 163(3), 370-378.

Schmit, J. M., Riley, M. A., Dalvi, A., Sahay, A., Schear, P. K., Schockley, K. D., and Pun, R. Y. K. (2006). Deterministic center of pressure patterns characterize postural instability in Parkinson's disease. *Experimental Brain Research,* 168(3), 357-367.

Seigle, B., Ramdani, S. and Bernard, P. L. (2009). Dynamical structure of center of pressure fluctuations in elderly people. *Gait and Posture,* 30(2), 223-226.

Shumway-Cook, A. and Woollacott, M. H. (1995). *Motor Control: Theory and Practical Applications.* Baltimore: Lippincott Williams and Williams.

Stanley, H. E., Amaral, L. A. N., Goldberger, A. L., Havlin, S., Ivanov, P. C., and Peng, C.-K. (1999). Statistical physics and physiology: Monofractal and multifractal approaches. *Physica A,* 270, 309-324.

Stergiou, N. and Decker, L. M. (2011). Human movement variability, nonlinear dynamics, and pathology: Is there a connection. *Human Movement Science,* 30(5), 869-888.

Stergiou, N., Harbourne, R. T. and Cavanaugh, J. T. (2006). Optimal movement variability: A new theoretical perspective for neurologic physical therapy. *Journal of Neurologic Physical Therapy,* 30, 120-129.

Stolze, H., Kuhtz-Buschbeck, J. P., Mondwurf, C., Boczeck-Funcke, A., Johnk, K., Deuschl, G., and Illert, M. (1997). Gait analysis during treadmill and overground locomotion in children and adults. *Electroencephalography and Clinical Neurophysiology,* 105, 490-497.

Takens, F. (1981). Detecting strange attractors in turbulence Vol. 898. D. Rand and L. S. Young (Eds.), *Dynamical Systems and Turbulence*

Trulla, L. L., Giuliani, A., Zbilut, J. P., and Webber, C. L. (1996). Recurrence quantification analysis of the logistic equation with transients. *Physics Letters A,* 223, 255-260.

Vaillancourt, D. E. and Newell, K. M. (2000). The dynamics of resting and postural tremor in Parkinson's disease. *Clinical Neurophysiology* 111(11), 2046-2056.

Vaillancourt, D. E. and Newell, K. M. (2002). Changing complexity in human behavior and physiology through aging and disease. *Neurobiology of Aging* 23(1), 1-11.

Van Orden, G. C., Holden, J. G. and Turvey, M. T. (2003). Self-organization of cognitive performance. *Journal of experimental psychology: general,* 132(3), 331-350.

Van Orden, G. C., Kloos, H. and Wallot, S. (2011). Living in the pink: Intentionality, wellbeing, and complexity. In: C. A. Hooker (Ed.), *Philosophy of Complex Systems. Handbook of the Philosophy of Science* (pp. 639-684). Amsterdam: Elsevier.

Vierordt, K. V. (1881). *Ueber das Gehen des Menschen in Gesunden und Kranken Zustaenden nach Selb-stregistrireden Methoden*. Tubingen: Laupp.

Von Porat, A., Henriksson, M., Holmstrom, E., Thorstensson, C. A., Mattsson, L., and Roos, E. M. (2006). Knee kinematics and kinetics during gait, step and hop in males with a 16 years old ACL injury compared with matched controls. *Knee Surgery, Sports Traumatology, Arthroscopy,* 14, 546-554.

Webber, C. L. and Zbilut, J. P. (1994). Dynamical assessment of physiological systems and states using recurrence plot strategies. *Journal of Applied Physiology,* 76(2), 965-973.

Webber, C. L. and Zbilut, J. P. (2005). Recurrence quantification analysis of nonlinear dynamical systems M. A. Riley and G. C. V. Orden (Eds.), *Tutorials in contemporary nonlinear methods for the behavioral sciences* Retrieved from http://www.nsf.gov/sbe/bcs/pac/nmbs/nmbs.jsp

West, B. J. and Goldberger, A. L. (1987). Physiology in fractal dimensions. *American Scientist,* 75, 354-365.

West, B. J. and Griffin, L. (1999). Allometric control, inverse power laws and human gait. *Chaos, Solitions and Fractals,* 10(9), 1519-1527.

West, B. J. and Scafetta, N. (2003). A nonlinear dynamical model of human gait. *Physical Review E,* 67, 051917.

Williams, G. P. (1997). *Chaos theory tamed.* Washington DC: John Henry Press.

Wrisley, D. M., Marchetti, G. F., Kurharsky, D. K., and Whitney, S. L. (2004). Reliability, internal consistency, and validity of data obtained with the Functional Gait Assessment, *Physical Therapy,* 84(10), 906-918.

Yakhdani, H. R. F., H. A. Bafghi, Meijer, O. G., Bruijn, S. M., Dikkenberg, N. v. d., Stibbe, A. B., ... Dieen, J. H. v. (2012). Stability and variability of knee kinematics during gait in knee osteoarthritis before and after replacement surgery. *Clinical Biomechanics,* 25, 230-236.

Zbilut, J. P. and Webber, C. L. (2008). Laminar recurrences, maxline, unstable singularities and biological dynamics. *The European Physical Journal-Special Topics* 164(1), 55-65.

In: Gait Biometrics
Editors: Li Li and Matthew Holmes

ISBN: 978-1-62948-381-8

Chapter IV

Biomechanical and Physical Activity Considerations for Diabetic Peripheral Neuropathy

Ryan T. Crews* and Sai V. Yalla

Center for Lower Extremity Ambulatory Research (CLEAR) at Scholl College of Podiatric Medicine at Rosalind Franklin University of Medicine and Science, North Chicago, IL, US

Abstract

Diabetic peripheral neuropathy (DPN) is a degradation in the nerves that transmit sensory data from the body (the lower extremities are most commonly effected) to the central nervous system. This most common type of neuropathy associated with diabetes was observed to develop in approximately 50% of individuals with diabetes. DPN has been shown to alter both gait as well as balance and also plays a key role in the development of diabetic foot ulcers that generally develop due to interplay between DPN and weight bearing activities. Although there is presently no definitive treatment for DPN there are some techniques and interventions that can be performed to mitigate the effects of DPN.

DPN causes a reduction in the ability to feel pressure, particularly in the feet. It also can also reduce the proprioception of the lower extremities. With this diminished sensory function it should not be surprising that balance during standing is compromised by DPN. The reduced sensory function not only resulted in increased postural sway but also higher ground reaction forces and variations of stride length during gait.

Moreover, infections that develop in diabetic foot ulcers are the primary cause of the extremely high amputation rate in individuals with diabetes. Physician and author Paul Brand noted DPN plays a key role in the development of ulcers by removing "the gift of pain." The loss of sensation in the feet allows individuals to continually physically

* Corresponding author: Ryan T. Crews, Center for Lower Extremity Ambulatory Research (CLEAR) at Scholl College of Podiatric Medicine at Rosalind Franklin University of Medicine and Science, 3333 Green Bay Rd. North Chicago, IL 60064. E-mail: Ryan.Crews@rosalindfranklin.edu.

damage their skin during weight bearing activities without inducing any protective reactions.

When considering the relationship between physical activity and diabetic foot ulcers, it appears the biggest concern is not the volume of physical activity one engages in. Instead, the day to day variability appears to be of highest importance.

Although presently there is no medication or treatments widely available to "cure" DPN, weight bearing exercises and balance training have been associated with positive improvement in postural control while not increasing risk of ulceration.

Introduction

Globally, the Diabetes Mellitus (DM) population is on the rise (Campbell and Branch, 1996; Karvonen, Tuomilehto, Libman, and LaPorte, 1993; Odden et al., 2011; Zimmet, McCarty, and de Courten, 1997) and is coupled with increased medical costs (Amos, McCarty, and Zimmet, 1997; Hex, Bartlett, Wright, Taylor, and Varley, 2012; Wolf and Colditz, 2012). Nerve disorders are a frequent occurrence in the DM population, out of which Diabetic Peripheral Neuropathy (DPN) is one of the most common (A. J. M. Boulton et al., 2005) in contrast to autonomic, proximal and focal types of neuropathies. In fact, in patients with a 20 or more year history of DM, the prevalence of DPN is 50% (Corriveau, Prince, Hebert, et al., 2000; National Diabetes Statistics, 2011).

In comparison to age and sex matched DM controls, individuals with diabetic neuropathy have significantly more comorbid medical conditions and higher healthcare charges and utilization primarily due to greater need of inpatient services (Zhao et al., 2009). Even within the DPN population, a 6-10% increase in medical costs has been observed as the severity of neuropathy increases (Currie et al., 2007; Shearer, Scuffham, Gordois, and Oglesby, 2003). In the US alone, the annual costs associated with DPN during 2001 was around 10.9 billion dollars (Gordois, Scuffham, Shearer, Oglesby, and Tobian, 2003). Similar increases of direct and indirect medical costs were found in other countries as well, and DPN has become a global economic burden (A. M. Barrett et al., 2007; A. J. M. Boulton, Vileikyte, Ragnarson-Tennvall, and Apelqvist, 2005). As it is evident that DPN is highly prevalent among the diabetic population (Tesfaye et al., 1996; Young, Boulton, MacLeod, Williams, and Sonksen, 1993), multiple cross sectional studies have been performed to understand the risk factors that influence the development of DPN among the DM population. Table 1 provides a list of studies which identified the prevalence of DPN in the DM population along with risk factors associated. Height (Adler et al., 1997; Katulanda et al., 2012), BMI, cardiovascular disease, blood pressure (Tesfaye, et al., 1996), and race (Gregg et al., 2004; L.A. Lavery, Armstrong, Wunderlich, Tredwell, and Boulton, 2003) were some of the risk factors that were identified, but not as common as the ones listed in Table 1. It should also be noted that DPN was identified by various clinical examinations such as Vibration Perception Threshold (VPT) tests and Mono Filament tests (MF) along with scoring such as Neuropathic Disability Score (NDS) and Neuropathy System Score (NSS). In one of the studies, a Neuropen and DN4 questionnaire was used (Table 1). The reported prevalence of DPN has ranged from 10.7% to 81%. Based on the total population and reported prevalence rates from each study, the mean prevalence was 35%. As it is evident from the table, age and duration of diabetes were the most common risk factors identified.

Table 1. Prevalence of DPN among multiple cross sectional studies on Diabetes Mellitus (DM) populations and identified risk factors

Study	Population	No of Participants	Examination	Prevalance	Risk Factors					
					Age	Duration	HbA1C	Prior ulceration	Smoking	Retinopathy
A. J. M. Boulton, Knight, Drury, and Ward, 1985	DM	382	MCV	10.70%	√	√				
Young, et al., 1993	DM	6487	VPT	28.50%	√	√	√			
Partanen et al., 1995	NIDDM	133	VPT	41.90%		√	√			
Tesfaye, et al., 1996	DM	3250	VPT	28%	√	√			√	√
Adler, et al., 1997	DM	775	MF	81%	√	√	√	√	√	
L. A. Lavery, et al., 2003	DM	1666	MF	41.40%				√		
Gregg, et al., 2004	DM	419	MF	28.50%	√					
Tesfaye et al., 2005	DM	1819	VPT	28%	√	√			√	√
Kumar et al., 2009	NIDDM	811	NDS	41.60%	√	√		√		
Van Acker et al., 2009	DM	1111	Neuropen, DN4	43%	√					
Mørkrid, Ali, and Hussain, 2010	DM	294	NSS, NDS	19.70%	√	√	√			
Tabatabaei-Malazy, Mohajeri-Tehrani, Madani, Heshmat, and Larijani, 2011	DM	124	MF	42%	√	√	√			
Katulanda, et al., 2012	DM	528	DNS	48.10%		√	√	√	√	√

DM - Diabetes Mellitus; NIDDM - Non Insulin Dependent Diabetes Mellitus; MCV - Mean Corpuscular volume; VPT - Vibration Perception Threshold; MF - MonoFilament; NDS - Neuropathic Disability Score; NSS - Neuropathy System Score; HbA1c - glycosylated haemoglobin

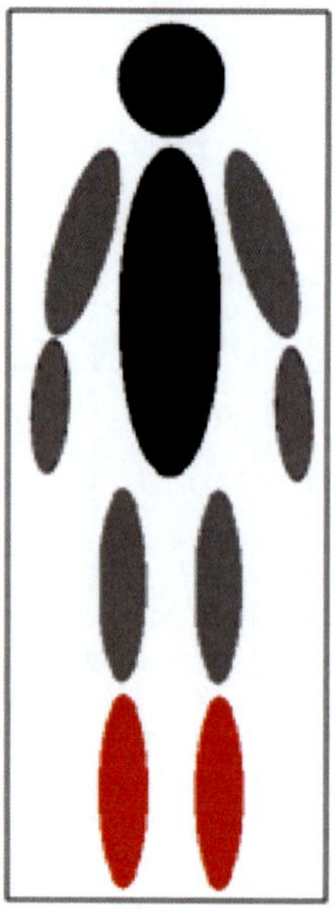

Figure 1. Toes and feet (highlighted in red) are the most commonly affected by DPN. Legs, hands, arms are also affected in some severe cases.

With the current aging population (Werner, 2011), DPN becomes a major concern in the growing elderly DM population. Elevated glycosylated dhaemoglobin (HbA1c) levels, prior ulceration, retinopathy and smoking were also found to be associated with the development of DPN. Although, risk factors such as HbA1c levels, ulceration, and smoking can be mitigated/ controlled with proper care and medication; duration and age are risk factors that cannot be controlled.

Although a number of detrimental changes to the nervous system are associated with DPN, this chapter will be referring to the most common type which is damage to the large nerve fibers that results in decreased sensorimotor function when referencing DPN (Tesfaye, 2006).

In DPN patients, peripheral nerves fails to transmit sensory data to the central nervous system due to the damage to the primary sensory neurons (McHugh and McHugh, 2004). Damage to the sensory neurons results in loss of sensation and numbness towards pain or temperature. The toes and feet are the major and most common areas of the human body affected by DPN along with arms, hands and legs (A. Boulton, Gries, and Jervell, 2004; A. J. M. Boulton, Vinik, et al., 2005).

DPN and Postural Stability

The human body needs to establish equilibrium between the nervous system and the muscular system in order to maintain postural stability during activities like standing or walking. The equilibrium is established through muscular contractions conducted in reaction to visual and sensory feedback (Adams, Gopher, and Lintern, 1975; Pearson, 2004). Visual and sensory feedback are used to assess the sway of the human body and help the brain to activate the necessary muscles to maintain stability.

Since the feet serve as the base of support and generally provide substantial sensory feedback during activities such as standing and walking, it is not surprising that DPN patients whom have diminished feedback from their feet have been observed to have increased sway during such activities (Menz, Lord, St George, and Fitzpatrick, 2004; Najafi et al., 2010). The ability to control this sway while standing has also been observed to be reduced significantly in diabetic patients with neuropathy. Similarly a pattern of increased center of mass sway has also been observed during gait in DPN patients (Menz, et al., 2004). Lack of postural control along with somnolence due to medication (Backonja, 2003) might result in higher risk of falls, increased depressive symptoms (Vileikyte et al., 2005), as well as reduced daily activates in ambulatory DPN patients.

Measuring postural sway is a common means of assessing postural stability. Table 2 provides multiple case controlled studies which measured postural sway with different types of equipment during standing. Although different types of equipment was used, it is evident from the table that DPN patients show increased postural sway when compared to the non-DPN diabetic population or healthy individuals. The percentage increase of sway in DPN patients when compared to healthy individuals ranged from 18% to 98%. Even when compared to the non-DPN diabetic population, DPN patients' postural instability increased from 57% to 117% based on the condition of testing. The percentages reported are net increase of instability when tested during eyes open and eyes closed conditions. Extensive literature review (RW Van Deursen and Simoneau, 1999; Wrobel and Najafi, 2010) on postural stability studies further validates the importance of visual and sensory feedback and their effect of DPN patients.

DPN patients need to develop some type of compensation in order to counter the increased postural sway (table 2). Lack in development of such compensation might lead to falls as a recent prospective cohort study on 9249 individuals aged above 67 years, showed higher incidence rates in people suffering from peripheral neuropathy (A.V. Schwartz et al., 2002). It was also observed that reaching distance while standing was significantly affected by peripheral neuropathy (Lin, Chen, Liao, and Chou, 2010). Peripheral neuropathy has been identified as a risk factor (Fulk, Robinson, Mondal, Storey, and Hollister, 2010; Macgilchrist et al., 2010) contributing to falls, which might not only be due to increased sway but also due to diminished strength in the muscles (Giacomozzi, D'Ambrogi, Cesinaro, Macellari, and Uccioli, 2008; Kruse, Lemaster, and Madsen, 2010; A. V. Schwartz et al., 2002). The equilibrium between the nervous system and muscular system is further hampered in DPN subjects with decreased muscle strength.

Decreased foot plantar flexion muscle strength (Macgilchrist, et al., 2010; Perlmutter, Lin, and Makhsous, 2010) has been observed in DM populations. Moreover, decreased ankle mobility and power were observed in DPN patients when compared to non-DPN diabetic patients (Mueller, Minor, Sahrmann, Schaaf, and Strube, 1994). Reduced muscle strength can especially effect the propulsive phase of the gait cycle which results in reduced walking speeds (Allet et al., 2010; Alvarenga, Pereira, and Anjos, 2010). Recent comprehensive review of literature (RW Van Deursen and Simoneau, 1999; Wrobel and Najafi, 2010) suggests significantly reduced walking speeds and gait alterations in DPN population when compared to DM and healthy individuals. The range of motion at knee has been seen to be reduced (Dingwell, Cusumano, Sternad, and Cavanagh, 2000; Meier, Desrosiers, Bourassa, and Blaszczyk, 2001) in DPN subjects resulting in higher stance times and double support times (Sacco and Amadio, 2000; Wrobel and Najafi, 2010). With increased double support times (Friedman, Munoz, West, Rubin, and Fried, 2002; Nordin, Moe-Nilssen, Ramnemark, and Lundin-Olsson, 2010; Winter, Patla, Frank, and Walt, 1990) combined with lack of muscle strength and increased postural sway (Allet et al., 2009), the DPN population is at very high risk of falls.

Table 2. Case controlled studies comparing postural stability in Healthy (H), Diabetes Mellitus (DM), and Diabetic Peripheral Neuropathy (DPN) patients with different types of equipment. All studies show increased postural instability in DPN patients

Study	Population	No of Participants	Equipment	Postural Instability % inc	
				DPN vs H	DPN vs DM
Simoneau, Ulbrecht, Derr, Becker, and Cavanagh, 1994	H, DM, DPN	51	Dynamometer, force platform	-	66-117%
J. K. Richardson and Hurvitz, 1995	H, DPN	40	Questionnaire	77%	-
Uccioli et al., 1997	DM, DPN	56	Standard Platform	60%	57%
Sacco and Amadio, 2000	H, DM, DPN	36	F scan system	-	-
Corriveau, Prince, Hébert, et al., 2000	H, DPN	30	Optotrak system, force plates	50%	-
Yamamoto et al., 2001	H, DM, DPN	210	Gravichart	-	-
Ducic, Short, and Dellon, 2004	DM, DPN	45	Matscan measurement system	-	87%
Lafond, Corriveau, and Prince, 2004	H, DPN	31	Force platform	30%	-
Manor and Li, 2009	H, DPN	24	Dynamometer, force plate, Motion analysis	18%	-
Najafi, Horn, et al., 2010	H, DPN	38	Body worn sensors	98%	-

Postural Stability Interventions/Techniques

In order to improve postural stability and prevent falls, the key aspect as mentioned earlier is to maintain the equilibrium between the nervous system and the muscular system. This can be achieved through

(a) Strength training for the muscles and
(b) Developing a control strategy for the postural sway (Tofthagen, Visovsky, and Berry, 2012)
(c) Enhance sensory feedback

Resistance/strength training in combination with balance/posture exercises have been shown to significantly reduce postural sway in the DM population (Morrison, Colberg, Mariano, Parson, and Vinik, 2010). Similar strength training (Kruse, et al., 2010), physiotherapeutic training (Allet, et al., 2010) and balance training (Salsabili, Bahrpeyma, Forogh, and Rajabali, 2011) using biodex stability system has been shown to improve postural control in the DPN population. More recently, training via a virtual reality concept using body worn sensors (Grewal et al., 2012) simulating obstacle crossing has also been shown to improve postural stability in DPN subjects. These simple and novel techniques develop the motor learning skills of the brain to achieve postural stability. This not only improves gait and balance in DPN population but also the quality of life. In addition to increasing muscular strength and improving postural control strategies, there has been some promising work in utilizing external stimulation to improve sensory feedback in elderly individuals.

Although much of the work has not been specific to DPN subjects, previous studies have found that the sensation threshold of tactile and proprioceptive systems is significantly lowered during electrical cutaneous stimulation (Collins et al., 2003; Dhruv, Niemi, Harry, Lipsitz, and Collins, 2002; Gravelle et al., 2002). The principle of these studies has been to apply sub-detection threshold (individuals can not feel the stimulus) neurological stimulation to the nerves in order to lower the threshold of stimulus required to cause the nerves to fire. In essence, the sub-detection threshold stimulus acts to boost the amplitude of secondary stimuli such as the application of pressure. One pilot study that did specifically look at DPN patients found that regular (5 times per week) provision of electrical stimulation via an aqueous solution resulted in a therapeutic benefit within their study sample. This randomized controlled study of 54 subjects found improvements in protective sensation and postural stability suggesting improved nerve function, however, the results have yet to be replicated in a large scale study (Najafi, Crews, and Wrobel, 2013).

The Association of DPN, Diabetic Foot Ulcers and Physical Activity

The lifetime incidence of developing a diabetic foot ulcer (DFU) has been estimated to be as high as 25% (Singh, Armstrong, and Lipsky, 2005), which makes it one of the most common of the complications associated with diabetes. As these wounds regularly act as a precursor to lower extremity amputation they also represent one of the most serious complications (A. J. M. Boulton and Vileikyte, 2000; Reiber et al., 2001). While there are many routes to limb amputation, most of them involve a wound, or foot ulcer (Pecoraro, Reiber, and Burgess, 1990). People with diabetes typically have impaired wound healing capacity, and as long as DFU remain open they present a portal for potentially limb threatening infections to enter the body.

The etiology of DFU in persons with diabetes is generally associated with the presence of DPN and cycles of repetitive stress generated by normal weight bearing activities such as walking (Brand, 1991).

As individuals with DPN have commonly lost protective sensation, or the "gift of pain," they are able to wear a hole in the surface of their feet with no more discomfort than a healthy person who wears a hole in their sock with repeated use. Stress to the foot is a product of the ground reaction force generated during weight-bearing activity. This force is composed of vertical, anteroposterior, and mediolateral components. The vertical force is much greater than the other two. At a fast walk, the vertical force amplitude reaches 1.5 times body weight (R. van Deursen, 2004a) and increases to 5 times body weight when running (Hamill, 2001). The vertical force damages healthy tissue by compressing and deforming it. Shear is the mechanical stress collectively generated by the anteroposterior and mediolateral components of ground reaction force. It causes tissue to tear by stretching it. During the foot loaded portion of gait as well as during periods of standing, the foot is exposed to moderate to high vertical compressive and shear stresses.

Additionally foot deformities, limited joint mobility, and partial foot amputations also often predispose the feet of diabetic individuals with DPN to areas of concentrated pressure and abnormal shear forces that significantly increase their risk of ulceration (A. J. Boulton, 1996; L. A. Lavery, Lavery, and Quebedeax-Farnham, 1995; L. A. Lavery, S. A. Vela, D. C. Lavery, and T. L. Quebedeaux, 1996). The pressure and shear stresses generated during activity lead to inflammation and eventually autolysis of the soft tissue beneath bony prominences of the foot (Brand, 1983, 1991).

Physical Activity Profiles of Individuals That Develop DFU

A number of studies have tried to gauge DFU risk and healing profiles based upon the amplitude of pressure (peak pressure) applied to the feet with each step as well as during static standing (Armstrong, Lavery, and Bushman, 1996; J. P. Barrett and Mooney, 1973; Duckworth, Boulton, and Betts, 1985). While higher magnitudes of plantar foot pressure have been associated with a concurrent increase in risk of ulceration in individuals with DPN (Frykberg et al., 1998; Stess, Jensen, and Mirmiran, 1997), researchers have been unable to identify a universal critical level of pressure that would predict future ulceration (Armstrong, Peters, Athanasiou, and Lavery, 1998).

Therefore it appears that not only the stress per step needs to be considered, but the time over which the foot is subjected to stress (such as the number of steps per day or week) should be considered in the etiologic formula for ulcer formation (Brand, 1991). With the knowledge that pressure alone does not provide sufficient knowledge to ascertain ulceration risk, researchers have begun to assess the quantity of physical activity individuals incur (Armstrong, Abu Rumman, Nixon, and Boulton, 2001; Armstrong et al., 2004; Maluf and Mueller, 2003). This has typically been done using accelerometers. Maluf and Mueller sought to investigate “cumulative plantar tissue stress” in subjects by multiplying the pressure time integral for a step (measured for patients in the laboratory) by the number of strides taken daily (Maluf and Mueller, 2003).

Thus by factoring in physical activity levels, researchers have tried to assess the cumulative trauma endured by diabetic feet. The fact that increased plantar pressure yields increased ulceration risk, suggests that increased activity levels (more cycles of stress) will also increase risk.

In their work evaluating the association of DFU and cumulative plantar tissue stress, Maluf and Mueller performed a cross-sectional study looking at the stress encountered by three different groups (Maluf and Mueller, 2003). Ten subjects were recruited for each of three groups. The groups consisted of 1) subjects with diagnosed diabetes mellitus and peripheral neuropathy and a history of recurrent ulcers, 2) subjects with diagnosed diabetes mellitus and peripheral neuropathy with no history of recurrent ulcers, and 3) non-diabetic control subjects. After the pressure time integral for a subject's step was measured, physical activity was monitored for seven days by an accelerometer. Contrary to expectations, subjects with diabetes and a history of recurrent ulcers did not have a greater level of activity and actually accumulated significantly less (41%) cumulative stress than the subjects with diabetes and no history of ulcers. Additionally, the subjects with diabetes and no history of recurrent ulcers had an equivalent level of daily cumulative stress as controls. Thus the individuals applying the least cumulative stress to their feet, were the ones who were developing ulcers.

In a similar natured investigation Armstrong et al., performed a prospective longitudinal study of activity level in persons at high risk of ulceration (Armstrong, et al., 2004). They also expected to find DFUs to be associated with higher levels of activity. One hundred subjects at high risk for diabetic foot ulceration were recruited. Risk factors included neuropathy, deformity, and/or a history of lower-extremity ulceration/partial foot amputation. Upon enrollment subjects were provided an accelerometer based activity monitor to be worn at the waist. Subjects donned the monitors continuously for a minimum of 25 weeks or until ulceration. Over the course of the study eight patients developed an ulcer, all of which were located on the plantar aspect of the forefoot. Again unexpected results were obtained when the ulcerated individuals demonstrated a significantly lower average daily activity level than the non-ulcerated subjects (809 ± 612.2 vs. 1394.5 ± 868.5 activity units).

In addition to the finding that subjects who ulcerated incurred less daily activity than those who did not, Armstrong et al., had a secondary finding of interest (Armstrong, et al., 2004). The group that ulcerated had significantly greater daily variability in activity. Furthermore in the two weeks prior to the development of ulcers, the ulcerative group's daily variability increased even more (CV =115.4 ± 43.0%) while the mean level of daily activity did not change. Thus, in addition to the amplitude of stress (pressure) and number of times the stress is applied, one should consider the time frame over which the applications of stress occur. It appears that ulcers are most likely to occur in individuals that are generally at the lower end of the activity spectrum, however, sporadically engage in bursts of activity. The greater variability in individuals that develop DFU helps to make sense of the counterintuitive activity data that shows individuals that ulcerate are the less active than those who do not.

With this perspective one may consider diabetic foot ulceration risk to increase in a similar manner as a "weekend warrior's" proneness to musculoskeletal injuries. The difference being that the weekend warrior likely goes from a low or moderate level of activity during the week to a high level during weekend sporting endeavors, while the patient with a diabetic ulcer goes from a low standard activity level to a brief period of moderate activity. Lott et al., previously presented a case study demonstrating this concept (Lott, Maluf, Sinacore, and Mueller, 2005). Approximately three weeks after having healed an ulcer, a patient donned an activity monitor in order to compare healed activity to the patient's activity level previously recorded during the healing process. The patient maintained a low level of activity on the first two days of monitoring, and then on the third day there was an approximate three fold spike in activity. The patient visited a county fair that day, and had re-ulcerated his foot by the end of the day. However, the cumulative stress generated that day was still less than the daily values previously recorded for subjects with diabetes, peripheral neuropathy, and no history of recurrent ulcers (Lott, et al., 2005; Maluf and Mueller, 2003).

While the initial studies looking at the role of physical activity in DFU formation have yielded good perspective, they have been limited in their assessments of activity. If they wanted a complete picture of daily physical activities, they had to rely on patient self report which is subject to both intentional and unintentional reporting errors. If they wanted objective measurements they had to rely on pedometers that were limited to measuring steps/day.

However recently it has become possible to unobtrusively assess both steps taken as well as other types of foot loading activities such as standing using a single wearable sensor (Aminian and Najafi, 2004; A. J. Boulton et al., 2008; Paraschiv-Ionescu, Buchser, Rutschmann, Najafi, and Aminian, 2004). Using such a sensor to monitor a more detailed analysis of physical activity of DPN patients revealed that the daily period of standing is almost twice the period of walking (Najafi, Crews, and Wrobel, 2010). On average DPN patients days were spent: 13.5±5.3% standing, 6.1±3.1% walking, 37.3 ±6.3% sitting, and 44.3±8.1 lying. The average total number of steps per day was 7,754± 4,087 and the number of episodes of continuous walking without stopping was 357±167 with maximum duration of 3.9±3.8 minutes or 422±403 steps. While the previous studies looking at the role of physical activity in DFU formation have generally been limited to the total number of steps per day, it appears that walking may cover as little as 3–13% of a person's daily physical activity. Therefore steps per day provide a rather limited representation of what the subject is doing during activities of daily living. This work highlights the importance of standing period as an important foot loading condition that requires further attention when treating and preventing patients with foot ulcers.

Association of Physical Activity to Healing and Preventing DFU

Offloading is the common term for the practice of reducing plantar pressure amplitudes. It is a key component in the treatment of DFU (A. J. M. Boulton, 2004; Cavanagh, Lipsky, Bradbury, and Botek, 2005; S.C. Wu, Crews, and Armstrong, 2005).

Conservative care for diabetic ulceration generally starts with external offloading designed to either prohibit loading of the foot or shift plantar pressure during gait from the wound area to an unaffected area (R. van Deursen, 2004b). Offloading techniques are based on the ambulation needs of patients with DFU and range from bed rest, wheelchairs, total contact casts, removable cast walkers, crutches, canes, in-shoe and ankle foot orthoses, and therapeutic shoes (Bus et al., 2008; S.C. Wu, et al., 2005). As the reduction in plantar pressure at the site of ulceration increases the chance for successful healing also increases (Bus, et al., 2008; Cavanagh and Bus, 2011; Knowles and Boulton, 1996). The best balance between patient ambulation and offloading is generally considered to be achieved with either removable cast walkers or total contact casts.

These types of footwear allow ambulation while providing offloading between the ranges of 64% to 92% compared to canvas rubber insole shoes (Cavanagh and Bus, 2011; L.A. Lavery, S.A. Vela, D.C. Lavery, and T.L. Quebedeaux, 1996). Despite findings that the best ulcer healing outcomes are accomplished with total contact casts (Bus, et al., 2008), they are not commonly used by actual practitioners (S. C. Wu, Jensen, Weber, Robinson, and Armstrong, 2008). The comfort and ease of application associated with removable cast walkers, both for the patient and the practitioner, make removable cast walkers the more commonly prescribed offloading footwear in ulcer treatment and might also make them more acceptable for current and future diabetic populations.

Studies have shown that removable cast walkers provide levels of plantar pressure reduction that are similar to those of total contact casts, (Baumhauer, Wervey, McWilliams, Harris, and Shereff, 1997; Gutekunst, Hastings, Bohnert, Strube, and Sinacore, 2011) though an increase in wound healing times has also been observed (Armstrong et al., 2001). The key contributor to the increased times needed for healing was compliance issues with removable cast walkers among ambulatory patients. Unfortunately the DPN that allows DFU to initially form also prevents subjects from feeling pain when walking on their wounds without the protection of an offloading device. This lack of pain can build false mental confidence that one's ulcerated feet are safe from injury, especially when patients are at home.

Another challenge to compliant use of offloaders is that they are traditionally cumbersome to walk in. Diminished plantar sensation, ulceration, and medications can all negatively affect postural stability during gait in diabetic patients with peripheral neuropathy (Menz, et al., 2004). Adding footwear that is difficult to walk in might further decrease postural stability, affecting a patient's willingness to comply, yet postural stability is often overlooked when making offloading design modifications.

Van Deursen (R. van Deursen, 2008) found that offloading devices negatively affect static and dynamic balance. Compared with total contact casts, removable cast walkers were found to have slightly better postural stability (L. Lavery et al., 1998). Considering that DPN puts individuals at risk of falls (J. K. Richardson, 2002) and that individuals with diabetic foot ulcers have been shown to have an even poorer balance than individuals with diabetic neuropathy alone, (Kanade, Van Deursen, Harding, and Price, 2008) it should come as no surprise that diabetic ulcer patients are reluctant to wear removable cast walkers that will further compromise their balance.

Challenging work remains to be completed regarding designing offloading devices that will allow wounds to heal, while also having limited detrimental impact on the capacity to perform activities of daily living (Crews, Sayeed, and Najafi, 2012).

Association of Physical Activity and Preventing DFU

The knowledge that those who ulcerate typically have low average physical activity levels and that the level of consistent daily weight-bearing activity does not increase DFU risk (Lemaster, Reiber, Smith, Heagerty, and Wallace, 2003) has led to the hypothesis that caregivers may be able to reduce their patients' risk for ulceration by helping them to gradually increase their activity level. In theory, at any given point in time each individual has a specific range of activity level that may be carried out safely. The individualistic nature of the "safe therapeutic range" is what has likely limited previous attempts to define universal values for DFU risk factors such as plantar pressure amplitude. Activity at a level greater than the safe range would result in injury, while a level below the safe range would result in weakening of the tissue and an increased risk for ulceration. The key to reducing ulceration risk through activity is for individuals to engage in an activity level at the upper portion of the safe range. This should result in restructuring and strengthening of the integument, including increased fibril diameters (Sanders, Goldstein, and Leotta, 1995; Wang and Sanders, 2003), without inducing an ulcer. The strengthening of the tissue should result in a shift of the activity continuum, allowing patients to safely engage in increased activity. Of equal importance to maintaining an appropriate mean daily activity level, will be to maintain limited variability from day to day. The occurrence of large bursts of activity must be prohibited in order to guard against ulceration.

References

Adams, J. A., Gopher, D. and Lintern, G. (1975). *The effects of visual and proprioceptive feedback on motor learning.* Paper presented at the Proceedings of the Human Factors and Ergonomics Society Annual Meeting.

Adler, A. I., Boyko, E. J., Ahroni, J. H., Stensel, V., Forsberg, R. C., and Smith, D. G. (1997). Risk factors for diabetic peripheral sensory neuropathy: results of the Seattle Prospective Diabetic Foot Study. *Diabetes care,* 20(7), 1162-1167.

Allet, L., Armand, S., de Bie, R. A., Golay, A., Monnin, D., Aminian, K., ... de Bruin, E. D. (2010). The gait and balance of patients with diabetes can be improved: a randomised controlled trial. *Diabetologia,* 53(3), 458-466. doi: 10.1007/s00125-009-1592-4

Allet, L., Armand, S., de Bie, R. A., Golay, A., Pataky, Z., Aminian, K., and de Bruin, E. D. (2009). Clinical factors associated with gait alterations in diabetic patients. *Diabet. Med.,* 26 (10), 1003-1009. doi: DME2811 [pii] 10.1111/j.1464-5491.2009.02811.x

Alvarenga, P. P., Pereira, D. S. and Anjos, D. M. (2010). Functional mobility and executive function in elderly diabetics and non-diabetics. *Rev. Bras. Fisioter.,* 14(6), 491-496. doi: S1413-35552010000600007 [pii]

Aminian, K. and Najafi, B. (2004). Capturing human motion using body-fixed sensors: Outdoor measurement and clinical applications. *Computer Animation and Virtual Worlds,* 15(2), 79-94.

Amos, A. F., McCarty, D. J. and Zimmet, P. (1997). The rising global burden of diabetes and its complications: estimates and projections to the year 2010. *Diabetic Medicine,* 14(S5), S7-S85.

Armstrong, D. G., Abu Rumman, P. L., Nixon, B. P., and Boulton, A. J. M. (2001). Continuous activity monitoring in persons at high risk for diabetes-related lower extremity amputation. *J. Am. Podiatr. Med. Assoc.,* 91, 451-455.

Armstrong, D. G., Lavery, L. A. and Bushman, T. R. (1996). Peak foot pressures influence healing time of diabetic ulcers treated with total contact casting. *Diabetologia,* 39 (Suppl. 1), A265.

Armstrong, D. G., Lavery, L. A., Holtz-Neiderer, K., Mohler, M. J., Wendel, C. S., Nixon, B. P., and Boulton, A. J. (2004). Variability in activity may precede diabetic foot ulceration. *Diabetes Care,* 27(8), 1980-1984.

Armstrong, D. G., Nguyen, H. C., Lavery, L. A., van Schie, C. H. M., Boulton, A. J. M., and Harkless, L. B. (2001). Off-loading the diabetic foot wound. *Diabetes care,* 24(6), 1019-1022.

Armstrong, D. G., Peters, E. J., Athanasiou, K. A., and Lavery, L. A. (1998). Is there a critical level of plantar foot pressure to identify patients at risk for neuropathic foot ulceration? *J. Foot Ankle Surg.,* 37(4), 303-307.

Backonja, M. (2003). Anticonvulsants for the treatment of neuropathic pain syndromes. *Current pain and headache reports,* 7(1), 39-42.

Barrett, A. M., Lucero, M. A., Le, T., Robinson, R. L., Dworkin, R. H., and Chappell, A. S. (2007). Epidemiology, public health burden, and treatment of diabetic peripheral neuropathic pain: a review. *Pain medicine,* 8(s2), S50-S62.

Barrett, J. P. and Mooney, V. (1973). Neuropathy and diabetic pressure lesions. *Orthop. Clin. North Am.,* 4, 43.

Baumhauer, J., Wervey, R., McWilliams, J., Harris, G., and Shereff, M. (1997). A comparison study of plantar foot pressure in a standardized shoe, total contact cast, and prefabricated pneumatic walking brace. *Foot and ankle international/American Orthopaedic Foot and Ankle Society [and] Swiss Foot and Ankle Society,* 18(1), 26.

Boulton, A., Gries, F. and Jervell, J. (2004). Guidelines for the diagnosis and outpatient management of diabetic peripheral neuropathy. *Diabetic Medicine,* 15(6), 508-514.

Boulton, A. J. (1996). The pathogenesis of diabetic foot problems: an overview. *Diabet. Med.,* 13(Suppl. 1), S12-16.

Boulton, A. J., Armstrong, D. G., Albert, S. F., Frykberg, R. G., Hellman, R., Kirkman, M. S., ... Wukich, D. K. (2008). Comprehensive foot examination and risk assessment: a report of the task force of the foot care interest group of the American Diabetes Association, with endorsement by the American Association of Clinical Endocrinologists. *Diabetes Care,* 31(8), 1679-1685. doi: 31/8/1679 [pii] 10.2337/dc08-9021

Boulton, A. J. M. (2004). Pressure and the diabetic foot: clinical science and offloading techniques. *The American journal of surgery,* 187(5), S17-S24.

Boulton, A. J. M., Knight, G., Drury, J., and Ward, J. D. (1985). The prevalence of symptomatic, diabetic neuropathy in an insulin-treated population. *Diabetes care,* 8(2), 125-128.

Boulton, A. J. M. and Vileikyte, L. (2000). Pathogenesis of diabetic foot ulceration and measurements of neuropathy. *Wounds,* 12 (Suppl. B), 12B-18B.

Boulton, A. J. M., Vileikyte, L., Ragnarson-Tennvall, G., and Apelqvist, J. (2005). The global burden of diabetic foot disease. *The Lancet,* 366(9498), 1719-1724.

Boulton, A. J. M., Vinik, A. I., Arezzo, J. C., Bril, V., Feldman, E. L., Freeman, R., ... Ziegler, D. (2005). Diabetic neuropathies a statement by the American Diabetes Association. *Diabetes care,* 28(4), 956-962.

Brand, P. W. (1983). The diabetic foot. In: Ellenberg, M., Rifkin, H. (Ed.), *Diabetes mellitus, theory and practice* (3rd ed., pp. 803-828). New York: Medical Examination Publishing.

Brand, P. W. (1991). The insensitive foot (including leprosy). In: Jahss, M. (Ed.), *Disorders of the Foot and Ankle* (2nd ed., pp. 2170-2175). Philadelphia: Saunders.

Bus, S., Valk, G., Van Deursen, R., Armstrong, D., Caravaggi, C., Hlaváček, P., ... Cavanagh, P. (2008). The effectiveness of footwear and offloading interventions to prevent and heal foot ulcers and reduce plantar pressure in diabetes: a systematic review. *Diabetes/ metabolism research and reviews,* 24(S1), S162-S180.

Campbell, P. R. and Branch, U. S. B. o. t. C. P. P. (1996). *Population projections for states by age, sex, race, and Hispanic origin: 1995 to 2025*: Population Projections Branch, Population Division, US Bureau of the Census.

Cavanagh, P. R. and Bus, S. A. (2011). Off-Loading the diabetic foot for ulcer prevention and healing. *Plastic and reconstructive surgery,* 127, 248S.

Cavanagh, P. R., Lipsky, B. A., Bradbury, A. W., and Botek, G. (2005). Treatment for diabetic foot ulcers. *The Lancet,* 366(9498), 1725-1735.

Collins, J. J., Priplata, A. A., Gravelle, D. C., Niemi, J., Harry, J., and Lipsitz, L. A. (2003). Noise-enhanced human sensorimotor function. *IEEE Eng. Med. Biol. Mag.,* 22(2), 76-83.

Corriveau, H., Prince, F., Hébert, R., Raîche, M., Tessier, D., Maheux, P., and Ardilouze, J. (2000). Evaluation of postural stability in elderly with diabetic neuropathy. *Diabetes care,* 23(8), 1187-1191.

Corriveau, H., Prince, F., Hebert, R., Raiche, M., Tessier, D., Maheux, P., and Ardilouze, J. L. (2000). Evaluation of postural stability in elderly with diabetic neuropathy. [Research Support, Non-U S Gov't]. *Diabetes Care,* 23(8), 1187-1191.

Crews, R. T., Sayeed, F. and Najafi, B. (2012). Impact of strut height on offloading capacity of removable cast walkers. *Clin. Biomech.* (Bristol, Avon). doi: S0268-0033(12)00045-9 [pii] 10.1016/j.clinbiomech.2012.03.001

Currie, C., Poole, C., Woehl, A., Morgan, C. L., Cawley, S., Rousculp, M., ... Peters, J. (2007). The financial costs of healthcare treatment for people with Type 1 or Type 2 diabetes in the UK with particular reference to differing severity of peripheral neuropathy. *Diabetic Medicine,* 24(2), 187-194.

Dhruv, N. T., Niemi, J. B., Harry, J. D., Lipsitz, L. A., and Collins, J. J. (2002). Enhancing tactile sensation in older adults with electrical noise stimulation. *Neuroreport,* 13(5), 597-600.

Dingwell, J., Cusumano, J., Sternad, D., and Cavanagh, P. (2000). Slower speeds in patients with diabetic neuropathy lead to improved local dynamic stability of continuous overground walking. *Journal of Biomechanics,* 33(10), 1269-1277.

Ducic, I., Short, K. W. and Dellon, A. L. (2004). Relationship between loss of pedal sensibility, balance, and falls in patients with peripheral neuropathy. *Annals of plastic surgery,* 52(6), 535-540.

Duckworth, T., Boulton, A. J. M. and Betts, R. P. (1985). Plantar pressure measurements and the prevention of ulceration in the diabetic foot. *Journal of Bone and Joint Surgery,* 67B, 79-85.

Friedman, S. M., Munoz, B., West, S. K., Rubin, G. S., and Fried, L. P. (2002). Falls and fear of falling: which comes first? A longitudinal prediction model suggests strategies for primary and secondary prevention. *Journal of the American Geriatrics Society,* 50(8), 1329-1335.

Frykberg, R. G., Lavery, L. A., Pham, H., Harvey, C., Harkless, L., and Veves, A. (1998). Role of neuropathy and high foot pressures in diabetic foot ulceration [In Process Citation]. *Diabetes Care,* 21(10), 1714-1719.

Fulk, G. D., Robinson, C. J., Mondal, S., Storey, C. M., and Hollister, A. M. (2010). The effects of diabetes and/or peripheral neuropathy in detecting short postural perturbations in mature adults. *Journal of neuroengineering and rehabilitation,* 7, 44-44.

Giacomozzi, C., D'Ambrogi, E., Cesinaro, S., Macellari, V., and Uccioli, L. (2008). Muscle performance and ankle joint mobility in long-term patients with diabetes. *BMC musculoskeletal disorders,* 9(1), 99.

Gordois, A., Scuffham, P., Shearer, A., Oglesby, A., and Tobian, J. A. (2003). The health care costs of diabetic peripheral neuropathy in the US. *Diabetes care,* 26(6), 1790-1795.

Gravelle, D. C., Laughton, C. A., Dhruv, N. T., Katdare, K. D., Niemi, J. B., Lipsitz, L. A., and Collins, J. J. (2002). Noise-enhanced balance control in older adults. *Neuroreport,* 13 (15), 1853-1856.

Gregg, E. W., Sorlie, P., Paulose-Ram, R., Gu, Q., Eberhardt, M. S., Wolz, M., ... Geiss, L. (2004). Prevalence of Lower-Extremity Disease in the US Adult Population≥ 40 Years of Age With and Without Diabetes 1999–2000 National Health and Nutrition Examination Survey. *Diabetes care,* 27(7), 1591-1597.

Grewal, G., Sayeed, R., Yeschek, S., Menzies, R. A., Talal, T. K., Lavery, L. A., ... Najafi, B. (2012). Virtualizing the Assessment: A Novel Pragmatic Paradigm to Evaluate Lower Extremity Joint Perception in Diabetes. *Gerontology,* 58(5), 463-471.

Gutekunst, D. J., Hastings, M. K., Bohnert, K. L., Strube, M. J., and Sinacore, D. R. (2011). Removable cast walker boots yield greater forefoot off-loading than total contact casts. *Clinical Biomechanics.*

Hamill, J., Hardin, E. C. (2001). Special Topics in Biomechanics. In: G. Kamen (Ed.), *Foundations of Exercise Science* (pp. 177-189). Baltimore: Lippincott Williams and Wilkins.

Hex, N., Bartlett, C., Wright, D., Taylor, M., and Varley, D. (2012). Estimating the current and future costs of type 1 and type 2 diabetes in the UK, including direct health costs and indirect societal and productivity costs. *Diabetic Medicine,* 29(7), 855-862.

Kanade, R., Van Deursen, R. W. M., Harding, K. G., and Price, P. (2008). Investigation of standing balance in patients with diabetic neuropathy at different stages of foot complications. *Clinical Biomechanics,* 23(9), 1183-1191.

Karvonen, M., Tuomilehto, J., Libman, I., and LaPorte, R. (1993). A review of the recent epidemiological data on the worldwide incidence of type 1 (insulin-dependent) diabetes mellitus. *Diabetologia,* 36(10), 883-892.

Katulanda, P., Ranasinghe, P., Jayawardena, R., Constantine, G. R., Sheriff, M. H. R., and Matthews, D. R. (2012). The prevalence, patterns and predictors of diabetic peripheral neuropathy in a developing country. *Diabetology and Metabolic Syndrome,* 4(1), 21.

Knowles, E. and Boulton, A. (1996). Do people with diabetes wear their prescribed footwear? *Diabetic Medicine,* 13(12), 1064-1068.

Kruse, R. L., Lemaster, J. W. and Madsen, R. W. (2010). Fall and balance outcomes after an intervention to promote leg strength, balance, and walking in people with diabetic peripheral neuropathy: "feet first" randomized controlled trial. *Phys. Ther.,* 90(11), 1568-1579. doi: ptj.20090362 [pii] 10.2522/ptj.20090362 [doi]

Kumar, S., Ashe, H., Parnell, L., Fernando, D., Tsigos, C., Young, R., ... Boulton, A. (2009). The Prevalence of Foot Ulceration and its Correlates in Type 2 Diabetic Patients: a Population-based Study. *Diabetic Medicine,* 11(5), 480-484.

Lafond, D., Corriveau, H. and Prince, F. (2004). Postural control mechanisms during quiet standing in patients with diabetic sensory neuropathy. *Diabetes care,* 27(1), 173-178.

Lavery, L., Fleishli, J., Laughlin, T., Vela, S., Lavery, D., and Armstrong, D. (1998). Is postural instability exacerbated by off-loading devices in high risk diabetics with foot ulcers? *Ostomy/wound management,* 44(1), 26.

Lavery, L. A., Armstrong, D. G., Wunderlich, R. P., Tredwell, J., and Boulton, A. J. M. (2003). Diabetic Foot Syndrome Evaluating the prevalence and incidence of foot pathology in Mexican Americans and non-Hispanic whites from a diabetes disease management cohort. *Diabetes care,* 26(5), 1435-1438.

Lavery, L. A., Lavery, D. C. and Quebedeax-Farnham, T. L. (1995). Increased foot pressures after great toe amputation in diabetes. *Diabetes Care,* 18(11), 1460-1462.

Lavery, L. A., Vela, S. A., Lavery, D. C., and Quebedeaux, T. L. (1996). Reducing dynamic foot pressures in high-risk diabetic subjects with foot ulcerations. A comparison of treatments. *Diabetes Care,* 19(8), 818-821.

Lavery, L. A., Vela, S. A., Lavery, D. C., and Quebedeaux, T. L. (1996). Reducing dynamic foot pressures in high-risk diabetic subjects with foot ulcerations: a comparison of treatments. *Diabetes care,* 19(8), 818-821.

Lemaster, J. W., Reiber, G. E., Smith, D. G., Heagerty, P. J., and Wallace, C. (2003). Daily weight-bearing activity does not increase the risk of diabetic foot ulcers. *Med. Sci. Sports Exerc.,* 35(7), 1093-1099. doi: 10.1249/01.MSS.0000074459.41029.75

Lin, S. I., Chen, Y. R., Liao, C. F., and Chou, C. W. (2010). Association between sensorimotor function and forward reach in patients with diabetes. *Gait Posture,* 32(4), 581-585. doi: S0966-6362(10)00230-4 [pii] 10.1016/j.gaitpost.2010.08.006 [doi]

Lott, D. J., Maluf, K. S., Sinacore, D. R., and Mueller, M. J. (2005). Relationship between changes in activity and plantar ulcer recurrence in a patient with diabetes mellitus. *Phys. Ther.,* 85(6), 579-588.

Macgilchrist, C., Paul, L., Ellis, B. M., Howe, T. E., Kennon, B., and Godwin, J. (2010). Lower-limb risk factors for falls in people with diabetes mellitus. *Diabet. Med.,* 27(2), 162-168. doi: DME2914 [pii] 10.1111/j.1464-5491.2009.02914.x

Maluf, K. S. and Mueller, M. J. (2003). Novel Award 2002. Comparison of physical activity and cumulative plantar tissue stress among subjects with and without diabetes mellitus and a history of recurrent plantar ulcers. *Clin. Biomech.* (Bristol, Avon), 18(7), 567-575.

Manor, B. and Li, L. (2009). Characteristics of functional gait among people with and without peripheral neuropathy. *Gait and posture,* 30(2), 253-256.

McHugh, J. M. and McHugh, W. B. (2004). Diabetes and peripheral sensory neurons: what we don't know and how it can hurt us. *AACN Advanced Critical Care,* 15(1), 136-149.

Meier, M. R., Desrosiers, J., Bourassa, P., and Blaszczyk, J. (2001). Effect of type II diabetic peripheral neuropathy on gait termination in the elderly. *Diabetologia,* 44(5), 585-592.

Menz, H. B., Lord, S. R., St George, R., and Fitzpatrick, R. C. (2004). Walking stability and sensorimotor function in older people with diabetic peripheral neuropathy1. *Archives of physical medicine and rehabilitation,* 85(2), 245-252.

Mørkrid, K., Ali, L. and Hussain, A. (2010). Risk factors and prevalence of diabetic peripheral neuropathy: A study of type 2 diabetic outpatients in Bangladesh. *International journal of diabetes in developing countries,* 30(1), 11.

Morrison, S., Colberg, S. R., Mariano, M., Parson, H. K., and Vinik, A. I. (2010). Balance training reduces falls risk in older individuals with type 2 diabetes. *Diabetes care,* 33(4), 748-750. doi: dc09-1699 [pii] 10.2337/dc09-1699 [doi]

Mueller, M. J., Minor, S. D., Sahrmann, S. A., Schaaf, J. A., and Strube, M. J. (1994). Differences in the gait characteristics of patients with diabetes and peripheral neuropathy compared with age-matched controls. *Physical therapy,* 74(4), 299-308.

Najafi, B., Crews, R.T. and Wrobel, J.S. (2013). A novel plantar stimulation technology for improving protective sensation and postural control in patients with diabetic peripheral neuropathy: a double-blinded, randomized study. *Gerontology,* 59(5), 473-480.

Najafi, B., Crews, R. T. and Wrobel, J. S. (2010). Importance of time spent standing for those at risk of diabetic foot ulceration. *Diabetes Care,* 33(11), 2448-2450. doi: dc10-1224 [pii] 10.2337/dc10-1224

Najafi, B., Horn, D., Marclay, S., Crews, R. T., Wu, S., and Wrobel, J. S. (2010). Assessing postural control and postural control strategy in diabetes patients using innovative and wearable technology. *J. Diabetes Sci. Technol.,* 4(4), 780-791.

NationalDiabetesStatistics, N. (2011) Retrieved Feb, 2012, from http://diabetes.niddk.nih.gov/dm/pubs/statistics/#fast

Nordin, E., Moe-Nilssen, R., Ramnemark, A., and Lundin-Olsson, L. (2010). Changes in step-width during dual-task walking predicts falls. *Gait and posture,* 32(1), 92-97.

Odden, M., Coxson, P., Moran, A., Lightwood, J., Goldman, L., and Bibbins-Domingo, K. (2011). The Impact of the Aging Population on Coronary Heart Disease in the United States. *The American journal of medicine.*

Paraschiv-Ionescu, A., Buchser, E. E., Rutschmann, B., Najafi, B., and Aminian, K. (2004). Ambulatory system for the quantitative and qualitative analysis of gait and posture in chronic pain patients treated with spinal cord stimulation. [Article]. *Gait and Posture,* 20 (2), 113-125.

Partanen, J., Niskanen, L., Lehtinen, J., Mervaala, E., Siitonen, O., and Uusitupa, M. (1995). Natural history of peripheral neuropathy in patients with non-insulin-dependent diabetes mellitus. *New England journal of medicine,* 333(2), 89-94.

Pearson, K. G. (2004). Generating the walking gait: role of sensory feedback. *Progress in brain research,* 143, 123-129.

Pecoraro, R. E., Reiber, G. E. and Burgess, E. M. (1990). Pathways to diabetic limb amputation: basis for prevention. *Diabetes Care,* 13, 513-521.

Perlmutter, S., Lin, F. and Makhsous, M. (2010). Quantitative analysis of static sitting posture in chronic stroke. *Gait and Posture,* 32(1), 53-56.

Reiber, G. E., Smith, D. G., Carter, J., Fotieo, G., Deery, H. G., 2nd, Sangeorzan, J. A., ... Boyko, E. J. (2001). A comparison of diabetic foot ulcer patients managed in VHA and non-VHA settings. *J. Rehabil. Res. Dev.,* 38(3), 309-317.

Richardson, J. K. (2002). Factors associated with falls in older patients with diffuse polyneuropathy. *J. Am. Geriatr. Soc.,* 50(11), 1767-1773. doi: jgs50503 [pii]

Richardson, J. K. and Hurvitz, E. A. (1995). Peripheral neuropathy: a true risk factor for falls. *The Journals of Gerontology Series A: Biological Sciences and Medical Sciences,* 50(4), M211.

Sacco, I. and Amadio, A. (2000). A study of biomechanical parameters in gait analysis and sensitive cronaxie of diabetic neuropathic patients. *Clinical biomechanics* (Bristol, Avon), 15(3), 196-202.

Salsabili, H., Bahrpeyma, F., Forogh, B., and Rajabali, S. (2011). Dynamic stability training improves standing balance control in neuropathic patients with type 2 diabetes. *J. Rehabil. Res. Dev.,* 48(7), 775-786.

Sanders, J. E., Goldstein, B. S. and Leotta, D. F. (1995). Skin response to mechanical stress: adaptation rather than breakdown--a review of the literature. *J. Rehabil. Res. Dev.,* 32(3), 214-226.

Schwartz, A. V., Hillier, T. A., Sellmeyer, D. E., Resnick, H. E., Gregg, E., Ensrud, K. E., ... Nevitt, M. C. (2002). Older Women With Diabetes Have a Higher Risk of Falls A prospective study. *Diabetes care,* 25(10), 1749-1754.

Schwartz, A. V., Hillier, T. A., Sellmeyer, D. E., Resnick, H. E., Gregg, E., Ensrud, K. E., ... Cummings, S. R. (2002). Older women with diabetes have a higher risk of falls: a prospective study. *Diabetes care,* 25(10), 1749-1754.

Shearer, A., Scuffham, P., Gordois, A., and Oglesby, A. (2003). Predicted costs and outcomes from reduced vibration detection in people with diabetes in the US. *Diabetes care,* 26(8), 2305-2310.

Simoneau, G. G., Ulbrecht, J. S., Derr, J. A., Becker, M. B., and Cavanagh, P. R. (1994). Postural instability in patients with diabetic sensory neuropathy. *Diabetes care,* 17(12), 1411-1421.

Singh, N., Armstrong, D. G. and Lipsky, B. A. (2005). Preventing foot ulcers in patients with diabetes. *Jama,* 293(2), 217-228.

Stess, R. M., Jensen, S. R. and Mirmiran, R. (1997). The role of dynamic plantar pressures in diabetic foot ulcers. *Diabetes Care,* 20(5), 855-858.

Tabatabaei-Malazy, O., Mohajeri-Tehrani, M., Madani, S., Heshmat, R., and Larijani, B. (2011). The Prevalence of Diabetic Peripheral Neuropathy and Related Factors. *Iranian Journal of Public Health,* 40(3).

Tesfaye, S. (2006). Diabetic Neuropathy, In: C. P. R. Boulton A. J. M., Rayman G. (Ed.), *The Foot in Diabetes* (4th ed.). West Sussex: John Wiley and Sons Ltd.

Tesfaye, S., Chaturvedi, N., Eaton, S. E. M., Ward, J. D., Manes, C., Ionescu-Tirgoviste, C., ... Fuller, J. H. (2005). Vascular risk factors and diabetic neuropathy. *New England journal of medicine,* 352(4), 341-350.

Tesfaye, S., Stevens, L., Stephenson, J., Fuller, J., Plater, M., Ionescu-Tirgoviste, C., ... Ward, J. (1996). Prevalence of diabetic peripheral neuropathy and its relation to glycaemic control and potential risk factors: the EURODIAB IDDM Complications Study. *Diabetologia,* 39(11), 1377-1384.

Tofthagen, C., Visovsky, C. and Berry, D. L. (2012). Strength and balance training for adults with peripheral neuropathy and high risk of fall: current evidence and implications for future research. *Oncol. Nurs. Forum,* 39(5), E416-424. doi: Y733N6573332RQ38 [pii] 10.1188/12.ONF.E416-E424 [doi]

Uccioli, L., Giacomini, P. G., Pasqualetti, P., Girolamo, S. D., Ferrigno, P., Monticone, G., ... Parisi, L. (1997). Contribution of central neuropathy to postural instability in IDDM patients with peripheral neuropathy. *Diabetes care,* 20(6), 929-934.

Van Acker, K., Bouhassira, D., De Bacquer, D., Weiss, S., Matthys, K., Raemen, H., ... Colin, I. (2009). Prevalence and impact on quality of life of peripheral neuropathy with or without neuropathic pain in type 1 and type 2 diabetic patients attending hospital outpatients clinics. *Diabetes and metabolism,* 35(3), 206-213.

Van Deursen, R. (2004a). Mechanical loading and off-loading of the plantar surface of the diabetic foot. *Clin. Infect. Dis.,* 39 Suppl. 2, S87-91.

Van Deursen, R. (2004b). Mechanical loading and off-loading of the plantar surface of the diabetic foot. *Clinical infectious diseases,* 39(Supplement 2), S87.

Van Deursen, R. (2008). Footwear for the neuropathic patient: offloading and stability. *Diabetes/metabolism research and reviews,* 24(S1), S96-S100.

Van Deursen, R. and Simoneau, G. (1999). Foot and ankle sensory neuropathy, proprioception, and postural stability. *J. Orthop. Sports Phys. Ther.,* 29(12), 718-726.

Vileikyte, L., Leventhal, H., Gonzalez, J. S., Peyrot, M., Rubin, R. R., Ulbrecht, J. S., ... Boulton, A. J. M. (2005). Diabetic peripheral neuropathy and depressive symptoms the association revisited. *Diabetes care,* 28(10), 2378-2383.

Wang, Y. N. and Sanders, J. E. (2003). How does skin adapt to repetitive mechanical stress to become load tolerant? *Med. Hypotheses,* 61(1), 29-35.

Werner, C. A. (2011). *The Older Population: 2010, 2010 Census Briefs*: United States Census Bureau.

Winter, D. A., Patla, A. E., Frank, J. S., and Walt, S. E. (1990). Biomechanical walking pattern changes in the fit and healthy elderly. *Physical therapy,* 70(6), 340-347.

Wolf, A. M. and Colditz, G. A. (2012). Current estimates of the economic cost of obesity in the United States. *Obesity research,* 6(2), 97-106.

Wrobel, J. S. and Najafi, B. (2010). Foot Technology, Part 1 of 2: Diabetic Foot Biomechanics and Gait Dysfunction. *Journal of diabetes science and technology,* 4(4), 833.

Wu, S. C., Crews, R. T. and Armstrong, D. G. (2005). The pivotal role of offloading in the management of neuropathic foot ulceration. *Current Diabetes Reports,* 5(6), 423-429.

Wu, S. C., Jensen, J. L., Weber, A. K., Robinson, D. E., and Armstrong, D. G. (2008). Use of pressure offloading devices in diabetic foot ulcers: do we practice what we preach? *Diabetes Care,* 31(11), 2118-2119.

Yamamoto, R., Kinoshita, T., Momoki, T., Arai, T., Okamura, A., Hirao, K., and Sekihara, H. (2001). Postural sway and diabetic peripheral neuropathy. *Diabetes research and clinical practice,* 52(3), 213-221.

Young, M., Boulton, A. J. M., MacLeod, A., Williams, D. R. R., and Sonksen, P. (1993). A multicentre study of the prevalence of diabetic peripheral neuropathy in the United Kingdom hospital clinic population. *Diabetologia,* 36(2), 150-154.

Zhao, Y., Ye, W., Le, T. K., Boye, K. S., Holcombe, J. H., and Swindle, R. (2009). Comparing clinical and economic characteristics between commercially-insured patients with diabetic neuropathy and demographically-matched diabetic controls. *Curr. Med. Res. Opin.,* 25(3), 585-597. doi: 10.1185/03007990802706289

Zimmet, P. Z., McCarty, D. J. and de Courten, M. P. (1997). The global epidemiology of non-insulin-dependent diabetes mellitus and the metabolic syndrome. *Journal of Diabetes and its Complications,* 11(2), 60.

In: Gait Biometrics
Editors: Li Li and Matthew Holmes
ISBN: 978-1-62948-381-8

Chapter V

Gait in Parkinson's Disease

Jan M. Hondzinski,[1,*] Zahra Kadivar[2] and Mia Frances B. Hegwood[1]
[1]Louisiana State University, Baton Rouge, LA, US
[2]Harris Health System, Houston, TX, US

Abstract

People with Parkinson's disease (PD) possess motor symptoms that influence several kinematic and kinetic aspects of gait. These deficits present particular problems as they place the person with PD at risk for falls and contribute to their perception of quality of life and disability. Although gait disturbances can occur during straight line walking they worsen in more challenging situations utilizing speed changes, for instance, gait involving turns, obstacle avoidance, and narrow corridors which further contribute to increased fall risk. However, a person with PD can improve their gait performance with training that supplements medical treatment. Utilization of repetitive practice, progressive exercise, and other motor learning principles result in physical benefits required when walking. Adopting the use of external cues such as auditory, visual, or tactile stimulation when training can also enhance motor performance even during challenging gait. Thus the alterations in walking gait biomechanics accompanying PD can improve by augmenting the standard medical treatment with training exercises which physically challenge and sensory stimulate the individual with PD.

This chapter does not offer a complete review of literature on the different aspects of Parkinson's gait. Instead, the chapter presents insight to various aspects of the disease to understand its influence on walking gait biomechanics. Sections include: (1) an introduction to PD; (2) clinical features of the disorder; (3) motor complications of PD gait; (4) non-motor influences on PD gait; (5) treatments for PD gait; and (6) a brief summary.

* Corresponding author: Jan M. Hondzinski, E-mail: jhondz1@lsu.edu.

Introduction

Definition and Etiology

Parkinson's Disease (PD) is a neurodegenerative disease which was first introduced by James Parkinson in the *Essay on the Shaking Palsy* in 1817 as "Paralysis agitans" (Vick, 1976). Known as the most common neurodegenerative disease after Alzheimer's disease, PD occurs in 13.4 of 10,000 Americans per year (Van Den Eeden et al., 2003). Estimates for the prevalence of Parkinson's disease include 0.3% of the entire population and 1% of adults over the age 60 in industrialized countries (Nussbaum and Ellis, 2003).

The main problems in PD follow dopaminergic deficiency of the basal ganglia (Tedroff, Ekesbo, Rydin, Langstrom, and Hagberg, 1999). Under normal conditions dopamine release allows the basal ganglia to serve as an internal trigger, enabling movements to occur in a sequential and smooth manner. Disruption of dopamine due to PD disrupts the normal functioning of the basal ganglia, including various voluntary movements such as those used in some aspects of gait (Morris, Iansek, Matyas, and Summers, 1994b) as well as autonomic functions of the gastrointestinal tract (Anderson et al., 2007). To better understand PD and its related gait deficits, one should first understand the normal functioning of the associated neural structures and pathways.

Major Nuclei and Connections of the Basal Ganglia

Nuclei. The basal ganglia (BG) involve a group of sub-cortical nuclei that have been grouped and sub-grouped various ways in the literature. For the purpose of this chapter BG composition will include: the striatum (STR), which includes the caudate, putamen and nucleus accumbens; the subthalamic nucleus (STN); the two substantia nigra (SN) divisions (pars reticulata-SNr and pars compacta-SNc), and the three globus pallidus (GP) divisions of the internal segment (GPi), external segment (GPe), and the ventral pallidum (Mink, 2007). Many BG nuclei inhibit or disinhibit their associated targets. The STR serve as the major input nuclei of the BG (Donoghue and Herkenham, 1986), thus contain several receptors for different types of neurotransmitters, most importantly dopamine. The STR receive input from different areas of the cerebral cortex (Monakow, Akert, and Kunzle, 1978), hippocampus, amygdala (Russchen, Bakst, Amaral, and Price, 1985), thalamus (Alexander, DeLong, and Strick, 1986) and SNc (Mink, 2007). The GPi and SNr act as the major output nuclei of BG (Nauta, 1979) to inhibit or release thalamic projections to the motor cortex (Alexander et al., 1986) needed to modulate movements.

Cortical connections. Numerous cortical area connections with the BG contribute to its functions. Research indicates projections for sensory (Hoover and Strick, 1993), cognitive, and motor (Middleton and Strick, 1997) information processing. Two motor cerebro-basal ganglia loops which start in the cerebral cortex, pass through the BG, and end back in the cerebral cortex exist (Albin, Young, and Penney, 1995; DeLong, 1990). The loops involve the direct and indirect pathways responsible for excitation and inhibition of voluntary movements, respectively, and are described briefly here (for greater details see (Alexander et al., 1986)).

In the direct pathway the STR inhibit the major output nuclei, removing their inhibitory effect on the thalamus, resulting in excitation of the cortex and facilitation of movements. In the indirect pathway the STR first inhibit the GPe, which removes the inhibition over the STN to facilitate the inhibitory effect of the output nuclei on the thalamus and suppress movement. More recently a third pathway was added to the classic two loop cerebro-basal ganglia model (Nambu, 2004; Nambu et al., 2000).

The relatively new hyperdirect pathway involves direct activation of STN from frontal lobe projections (Nambu et al., 2000) leading to movement inhibition similar to the indirect pathway, but faster (Nambu, 2004). During voluntary movement the three pathways work together to withhold (hyperdirect), release (direct), and terminate (indirect) the desired motor commands (Nambu, 2004) and contribute a great deal to the normal functioning of the BG (Obeso et al., 2000).

Sub-cortical connections. The basal ganglia directly and indirectly connect to sub-cortical structures. Projections of the BG connect to the mesencephalic locomotor region (MLR) (Mileykovskiy, Kiyashchenko, Kodama, Lai, and Siegel, 2000; Skinner and Garcia-Rill, 1984) and the pedunculopontine nucleus (PPN) (Kelland and Asdourian, 1989) of the mesopontine tegmentum (Shik, Severin, and Orlovskii, 1966) thus target the brainstem (Inglis et al., 1994; Inglis and Winn, 1995). The MLR projections through the medial reticulo-spinal tract, which stems from the midbrain region and passes through the ventromedial medulla and spinal cord (Rye, Lee, Saper, and Wainer, 1988), activate central pattern generators (CPG) in cats (Garcia-Rill, 1991) for locomotion control (Inglis and Winn, 1995). Stimulation of areas (e.g. PPN and SNr) with major associations to MLR evoke locomotion in rats (Skinner, Kinjo, Ishikawa, Biedermann, and Garcia-Rill, 1990) and cats (Lai and Siegel, 1990; Takakusaki, Habaguchi, Ohtinata-Sugimoto, Saitoh, and Sakamoto, 2003) provide further support for the involvement of MLR in gait. Although neurological structures of cats and rats are not identical to that of humans, scientists regard the PPN projections to the BG extremely vital to movement function (Mena-Segovia, Bolam, and Magill, 2004; Parent and Hazrati, 1995; Webster, 1990), however little research on CPG activity in people with PD exists.

The numerous individual roles of the BG nuclei act collectively to function as a whole. Thus, a lesion in one nucleus of the BG can affect whole system function (Vick, 1976).

Diagnosis of PD

Diagnosis of PD presently remains an inexact science. Symptoms of disorders such as secondary (acquired) parkinsonism, heredodegenerative parkinsonism, and multiple system degeneration mimic the major symptoms of tremors, rigidity, bradykinesia, and imbalance observed in PD, also known as idiopathic (primary) parkinsonism (Jankovic, 2008). Attenuation of aforementioned signs to levodopa therapy can help differentiate PD from other parkinsonian disorders (Jankovic, 2008) as well as other diseases with parkinsonism symptoms. However, failure of these criteria to diagnose some PD patients (Hughes, Daniel, Blankson, and Lees, 1993; Parati et al., 1993) led to development of other diagnostic tools. Currently, neuroimaging studies (Gorell et al., 1995; Yekhlef et al., 2003) add to the accuracy of clinical diagnosis of PD, however, further investigations are required to establish sensitivity of these tests before adoption as part of a standard diagnostic protocol.

Stages of PD

Two major disease severity scales, the Hoehn and Yahr (H&Y) and the Unified Parkinson's Disease Rating Scale (UPDRS), are currently used to assess the disease level and progression, symptom severity, and functional limitations of people with Parkinson's disease. Both are frequently used in research as selection criteria or study participation eligibility. Clinicians use the scales to monitor disease severity by identifying new symptom onset and tracking symptom progression.

Originally established in 1967 (Hoehn and Yahr, 1967), the modified H&Y evolved as the primary source in research and clinical assessment of functional disability and overall disease severity (Goetz et al., 2004). The stages range from 0 to 5 to indicate disease progression from no visible signs of the disease to affecting one and then both sides of the body. The stages also include the progression of balance impairment and the gradual loss of physical independence and mobility without assistance. Table 1 provides the detailed descriptions of the Modified H&Y stages.

The original UPDRS (Fahn et al., 1987) gained literary popularity in both clinical and research settings as the recent gold standard for PD severity ratings (Robichaud et al., 2005). This scale includes a rating from the Modified H&Y scale and similarly assesses disease severity. The UPDRS provides a comprehensive assessment of disease severity, covering functional impairment in addition to motor and non-motor symptoms. Separate rating scales, ranging from 0 = no, 1 = slight, 2 = mild, 3 = moderate, to 4 = severe impairment, developed for each section include: mentation, mood and behavior (Part I); activities of daily living (Part II); in depth motor examination of both upper and lower body movements (Part III); and motor complications (Part IV). The UPDRS-III includes a complete evaluation of the motor symptoms and the scoring offers a more comprehensive assessment of motor symptom severity than the H&Y scales.

Table 1. Modified H&Y stages of Parkinson's disease

Stage	Description
0	No disease symptoms
1	Unilateral symptoms
1.5	Unilateral symptoms plus axial involvement
2	Mild bilateral symptoms with unimpaired balance
2.5	Mild bilateral symptoms with recovery on pull test
3	Mild to moderate bilateral symptoms with postural instability and physical independence
4	Severe symptoms with unassisted walk and/or stand abilities
5	Severe symptoms causing immobility unless aided

Adapted from (Gancher, 2006).

Muscle rigidity, facial expressions and speech, finger and toe tapping, hand and wrist movements, the four cardinal motor symptoms, as well as certain activities of daily living motions are assessed in UPDRS-III, with a score range of 0 to 108 for no to severe impairment, respectively. In general, UPDRS-III scores less than or equal to 11.2 denote mild impairment and scores of 54.4 and greater denote severe impairment, while score changes greater than 5 points denote clinically significant changes in symptom severity (Schrag,

Sampaio, Counsell, and Poewe, 2006) that can become more severe with disease progression (Holford and Nutt, 2008; Morrish, Sawle, and Brooks, 1996) or less severe with training/rehabilitation (e.g., (Corcos et al., 2012)).

Critiques of the original UPDRS scale (Goetz, Poewe, Rascol, Sampaio, and Stebbins, 2003) resulted in the development of the Modified Unified Parkinson's Disease Rating Scale, or MDS-UPDRS to address ambiguities in the original, offer more straightforward instructions and new clinimetric evaluations, and introduce additional non-motor signs and symptoms (Goetz et al., 2008). Part I now includes daily non-motor living tasks, common sleep disturbances, and autonomic symptoms. Parts II and III remain very similar to the original and Part IV assesses motor complications. The MDS-UPDRS five-point scale correlates highly to the original scale yet may better emphasize early disease symptom changes and separate the slight and mild ratings from normal and severe ratings (Goetz et al., 2008), thus could eventually replace the original scale.

Clinical Features of Parkinson's Disease

Motor complications. PD persists primarily as a motor disorder even though people with the disease can possess several motor and non-motor symptoms. The motor disorder label stems from the four cardinal symptoms of PD which include tremor at rest, rigidity, bradykinesia (akinesia), and imbalance (postural instability) (Vick, 1976). In addition, flexed posture and freezing (motor blocks) have been included among classic features of parkinsonism (Pahwa, Lyons, Koller, and Jankovic, 2007). The symptom of imbalance results from the various motor complications that usually appear unilateral and gradually become more symmetric throughout disease progression (Vick, 1976). Table 2 provides definitions for each motor symptom observed in PD.

Non-Motor Complications

Motor symptoms associated with PD not only vary by individual, but also vary and progress in severity with progression of the disease as observed in stage advancement. In contrast, non-motor symptoms are not always observable and do not necessarily advance with disease progression. However, the significance of non-motor symptoms arises when they contribute to motor dysfunction or poor quality of life at the end stage of PD. This significance would increase if they eventually contribute to disease diagnosis.

Behavioral and cognitive symptoms. Clinically defined by motor dysfunctions, Parkinson's disease also encompasses non-motor behavioral symptoms such as mild cognitive impairment, dementia, depression, and anxiety. Decreased serotonin and dopamine levels exist in Parkinson's patients who suffer from depression or anxiety symptoms such as sadness, panic attacks, social avoidance, depressed mood, and pessimism (McDonald, Richard, and DeLong, 2003; Prediger, Matheus, Schwarzbold, Lima, and Vital, 2012). Mild cognitive impairment (MCI) in people with PD involves attentional and executive dysfunctions and in some cases other cognitive deficiencies in memory and learning (Kulisevsky, 2000). In fact people with PD and MCI risk early onset of dementia (Aarsland,

Bronnick, and Fladby, 2011). Recent literature reveals that behavioral disturbances, cognitive impairment, and the associated neurodegenerative processes can be detected much earlier in the disease process than when the cardinal motor symptoms appear (Watson et al., 2012). Unfortunately, behavior and cognitive impairments also occur in other disorders such as Alzheimer's and Huntington's diseases (Hodges, Salmon, and Butters, 1990), offering no current early diagnostic value of these symptoms in isolation for PD.

Table 2. Parkinson's disease motor symptoms and associated gait complications

Symptom	Definition	Clinical Gait Manifestation
Tremor	Involuntary rhythmic oscillation of at least one functional body region. *Resting tremor* occurs at rest. *Action tremor* occurs during slow body movements.	Foot oscillation during gait initiation and/or swing (Deuschl, Volkmann, and Raethjen, 2007)
Rigidity	A state of muscular inflexibility; stiff muscle tone of the affected limb which cannot easily relax	Reduction of deep tendon reflexes and postural deformities (Fung and Thompson, 2007)
Bradykinesia	Slowness of any voluntary movement; difficulty planning and initiating voluntary movement	Step and arm swing slows at initiation of and during walking (Rosin, Topka, and Dichgans, 1997); reduced limb coordination (Winogrodszka, Booji, and Wolters, 2005)
Hypokinesia	Reduction of movement size	Small step length and arm swing cycle (Rosin et al., 1997); altered regulation of stride length and increase in cadence (Morris et al., 1994b)
Akinesia	Difficulty with initiation of movement; impaired muscle activation of a voluntary movement	Trouble initiating walk sequence (Sellby, 1990)
Dyskinesia	Involuntary choreic movements which usually occurs secondary to medication (Harold, Klawans, and Barr, 1990)	Choreic, dystonic, or festinating gait, shuffling, and freezing of gait (Giladi, Treves, et al., 2001)

Note: Due to emphasis of this chapter on gait complications, only gait related clinical manifestations are reported.

Sensory symptoms. People with PD often deal with sensory deficits beyond those observed during normal aging. Alterations in somatosensations, vision, vestibular responses, olfaction, and pain prevalent in this population can occur prior to the more common motor deficits and/or manifest as motor dysfunctions, masking the actual affliction. A brief emphasis on some somatosensory, visual, and vestibular problems offers insight to the sensory dysfunctions that can ultimately influence PD gait.

Somatosensations of the skin, muscles, and joints may decline in those with PD. Skin biopsies reveal loss of epidermal nerve fibers and Meissner corpuscles, while Quantitative Sensory Testing reveals decreased abilities to perceive tactile and thermal sensations as well as mechanical pain perception (Nolano et al., 2008). Insufficient perceptions of proprioception (Almeida et al., 2005) and kinesthesia (Adamovich, Berkinblit, Hening, Sage, and Poizner, 2001) also appear in people with PD. While different opinions for the pathophysiologic causes of the sensory dysfunction exist, dopaminergic medications seem to worsen sensory perceptions of kinesthesia (Wright et al., 2010) and may worsen (O'Suilleabhain, Bullard, and Dewey, 2001) or improve (Almeida et al., 2005) proprioceptive sensations in this populace.

Visual problems in PD can, but do not have to, result from eye movement difficulties. Inadequate eye movement amplitudes using smooth pursuit eye movements make it difficult to visually track objects (Shibasaki, Tsuji, and Kuroiwa, 1979). Hypometric eye movement amplitudes (Crawford, Goodrich, Henderson, and Kennard, 1989) and increased latencies (White, Saint-Cyr, Tomlinson, and Sharpe, 1983) during saccadic eye movements make it difficult to reposition the eyes and often require a series of saccades to complete the task. Visual suppression, thus inadequate visual perception, occurs during performance of saccadic eye movements (Starr, Angel, and Yeates, 1969) thus often occurs more in the PD population. Other prevalent visual deficits in PD, not linked to eye movements, include those in identification and contrast perception as well as discrimination of color, orientation, and motion (Armstrong, 2011). Ironically, the medicines commonly used by those with PD to improve motor symptoms, mood, or visual problems can cause or contribute to visual deficits in PD (Armstrong, 2011).

Vestibular deficits also occur in people with PD. Investigators observed vestibular deficits in response to bithermal caloric tests, electronystagmography (Reichert, Doolittle, and McDowell, 1982), and sinusoidal rotations (Weinrich and Bhatia, 1986) for these individuals only when medicated (i.e., levodopa). Non-medicated people with PD produced reduced vestibular-evoked myogenic potential amplitudes suggesting abnormal saccule activity (Potter-Nerger, Reich, Colebatch, Deuschl, and Volkmann, 2012), yet normal vestibular and cervical perceptions of head and/or neck rotations (Nakamura, Bronstein, Lueck, Marsden, and Rudge, 1994).

These outcomes suggest that anti-Parkinson's medications induce vestibular problems in PD, and that the disease itself may only influence part of the peripheral vestibular apparatus. This would explain why results of vestibular chair testing, which stimulates all peripheral vestibular receptors in different ways, successfully identified most participants as controls or with PD (Dastgheib, Lithgow, and Moussavi, 2012).

Autonomic symptoms. Parkinson's related neurodegeneration can also exhibit dysfunction through autonomic nervous system (ANS) control (see Table 3). Although individual autonomic symptoms associated with Parkinson's disease are not deadly, concern exists for individuals with several symptoms which could result in respiratory or cardiovascular difficulties. For instance, sialorrhea could become problematic when paired with dysphagia, if an individual aspirates the saliva, causing respiratory arrest. Although ANS symptoms can vary greatly across individuals with PD, orthostatic hypotension presents in almost half this population (Allcock, Ullyart, Kenny, and Burn, 2004).

Table 3. Parkinson's disease autonomic symptoms

Autonomic Symptom	System Dysfunction
Orthostatic hypotension	Cardiovascular (Goldstein, 2006)
Resting heart rate variability decreases Deep breathing-respiratory sinus arrhythmia	Cardiovascular (Jain et al., 2011)
Sialorrhea (excessive saliva production) Dysphagia (trouble swallowing) Frequent nausea	Gastrointestinal (Pandya, Kubu, and Giroux, 2008)
Urinary urgency Constipation	Gastrointestinal (Sakakibara et al., 2011)
Impotence	Reproductive (Sakakibara et al., 2011)
Breathing difficulties Stridor (high pitched breathing)	Respiratory (Glass, Josephs, and Ahlskog, 2006)
Hyperhidrosis (excessive sweating)	Integumentary system (Pursiainen, Haapaniemi, Korpelainen, Sotaniemi, and Myllyla, 2007)

Motor Complications on Parkinson's Gait

Motor complications associated with PD affect body posture and body movements. Although walking depends directly on movement of the lower extremities, movements of the upper extremities, and trunk also contribute to normal or abnormal gait function. In fact, common contributors to imbalance and the abnormal multi-limb coordination observed in PD gait include muscular rigidity of limb joints and of the torso (Ueno, Yanagisawa, and Takami, 1993), the latter of which contributes to the forward leaning posture, prominent in this population.

The following text focuses on providing the reader insight to the classic gait complications associated with PD. A description of the kinematic and kinetic features associated with PD gait offer readers greater insight to the underlying causes of these complications.

Classic Gait Complications

Gait difficulties persist as one of the most noticeable motor deficits in individuals with PD (Murray, Sepic, Gardner, and Downs, 1978). Gait abnormalities yield to difficulties in gait initiation (akinesia) (Giladi, Balash, Ruzicka, and Jankovic, 2007; Sellby, 1990), increased freezing of gait, increased stride-to-stride variability (Blin, Ferrandez, and Serratrice, 1990), as well as complications with gait termination (Bishop, Brunt, and Marjama-Lyons, 2006).

Since people with PD possess a nine times greater risk of recurrent falls than that of age-matched controls (Bloem, Grimbergen, Cramer, Willemsen, and Zwinderman, 2001) and because approximately 80% of PD patients on medication show symptoms of freezing and akinesia (Sellby, 1990) gait deficits become a major concern to this population. Discussing the kinematic and kinetic alterations in various divisions of the PD gait cycle (gait initiation, progression, and termination) should help explain the overall biomechanics of PD gait.

Gait Biomechanics—Gait Initiation

Gait initiation involves a transition from a stationary double support position, to a dynamic gait cycle (Levangie and Norkin, 2001). Gait akinesia or "start hesitation," one landmark of Parkinson's gait (Sellby, 1990) observed especially after long periods of immobility, occurs when the person takes several short steps or step attempts before they can generate a large enough force that can lead to a normal step length (Giladi et al., 2007). Changes in gait initiation kinematics can be directly linked to the gait initiation kinetics in those with PD.

Kinematics. Gait initiation kinematics include the control of whole body equilibrium (Martin et al., 2002). Before initiating gait, one should attain a stable condition and control the location of the vertical height of the center of mass (COM) above the ground (posture) (Halliday, Winter, Frank, Patla, and Prince, 1998). Because attaining the stable condition demands more time and effort in those with PD and velocity and amplitude of forward COM displacement reduces significantly for the preparatory (postural) and stepping phases of gait initiation (Halliday et al., 1998), hesitation in starting the gait cycle develops (Gantchev, Viallet, Aurenty, and Massion, 1996).

Kinetics. Production of adequate force and accurate timing of force production must occur for accurate movement control. Generation of sufficient force through major lower limb muscles (Bronstein, Brandt, Woolacott, and Nutt, 2004; Levangie and Norkin, 2001) contribute to kinetics of gait initiation. Abnormal force control, evident in people with PD, becomes more significant in patients at later stages of the disease (Mitoma, Hayashi, Yanagisawa, and Tsukagoshi, 2000). Slower gait initiation can result from improper force production in the plantar and dorsi flexors due to a stooped posture (Crenna and Frigo, 1991), in the antigravity muscles of the leg due to impairment in the sensory detection of body load (Dietz, Leenders, and Colombo, 1997), and in other homologous muscles due to central mechanisms (Gantchev et al., 1996) like the supplementary motor area (SMA). In a healthy individual activation of the SMA precedes gait initiation, reflecting its role in planning different components of gait (Crenna and Frigo, 1991). Termination of SMA activity by signals from the BG result in activity in the primary motor cortex (M1) and preparation for planning of the next sub movement (Georgiou et al., 1993; Meara and Koller, 2000). Therefore, if the BG fail to trigger SMA on time, the activation of M1, the execution of the current movement (e.g. gait initiation), and the preparation of the subsequent movement (gait progression) do not occur (Georgiou et al., 1993; Meara and Koller, 2000).

Abnormal kinetic activity and its detection lead to abnormal gait initiation kinematics in PD. Inadequate detection of gravitational forces leads to difficulties in postural adjustments (Dietz et al., 1997), thus delays the preparatory phase, while the BG delay in activating SMA leads to delays in recruiting motor units (Crenna and Frigo, 1991) needed to produce propulsive forces during the stepping phase. At best, the phase delays and insufficient anteriorly directed force production by the supporting/propelling foot, result in a slower and smaller step during gait initiation (Gantchev et al., 1996).

Gait Biomechanics—Gait Progression

Gait patterns within gait progression commonly adhere to a rhythmic and cyclic pattern in the normal functioning adult. Although this activity differs from gait initiation, the abnormalities due to PD persist. The following sections will correspond to major motor complications associated with gait progression, including the kinematics and kinetics of walking and freezing of gait.

Kinematics. Stooped posture, shuffling of feet, and reduction of arm swing and joint range of motion (ROM) (Merritt, 1963), primary manifestations of Parkinson's gait, cause PD patients to walk with flexed knees, hips, and ankles, shorter stride length, and support more weight on their forefoot (Sellby, 1990; Vieregge, Stolze, Klein, and Heberlein, 1997). Shuffling of gait results from severe weakness in ankle dorsi flexors, causing dangling toes during the swing phase, known as foot drop (Mitoma et al., 2000). Shuffling of gait makes it difficult to discern the heel contact and toe off phases of the gait cycle and remains the primary cause of PD patients' inability to clear obstacles during gait progression, thus contributing to falls (Bloem, Hausdorff, Visser, and Giladi, 2004). Additional characteristics of PD gait include decreased step amplitude (Blin et al., 1990; Morris, 2000) and gait speed (Hausdorff, Cudkowicz, Firtion, Wei, and Goldberger, 1998; Pedersen, Oberg, Larsson, and Lindval, 1997), and increased cadences to compensate for the shorter stride length resulting in increased step numbers (Morris, Iansek, Matyas, and Summers, 1994a). Limb stiffness, tremor, and bradykinsia may explain greater arm swing asymmetry during PD gait, while tremor may explain less upper limb coordination (Huang et al., 2012).

Stride-to-stride variability during walking reflects the important rhythmic feature of gait (Hausdorff et al., 1998). Arrhythmia describes increase in stride-to-stride variability (Schaafsma et al., 2003), where rather than a continuous sequential movement, the individual performs a series of disconnected strides (Hausdorff et al., 1998). Temporal and spatial variability of certain gait parameters remain constant (coefficient of variation for stride length and stride time remain between 3-6% and less than 3%, respectively) (Menz, Latt, Tiedemann, Mun San Kwan, and Lord, 2004) as one ages, unless some type of complication occurs (Gabell and Nayak, 1984; Hausdorff et al., 1997). Therefore, alterations in variability of gait can reflect a pathological dysfunction (Hausdorff et al., 1997).

People with PD show increased variability of spatial and temporal components of gait (Blin et al., 1990; Hausdorff et al., 1998; MacKay-Lyons, 1998). Increases in variability of gait cycle duration (Baltadjieva, Giladi, Gruendlinger, Peretz, and Hausdorff, 2006; Hausdorff et al., 1998; MacKay-Lyons, 1998) and stride length (Blin, et al., 1990; Lewis, Byblow, and Walt, 2000; MacKay-Lyons, 1998) exist at very early stages of PD (Schaafsma, et al., 2003) and precede changes in step velocity and length (Baltadjieva et al., 2006). Variability in stride length and stride duration significantly correlate with frequency of falls and episodes of freezing in PD (Hausdorff, Rios, and Edelberg, 2001; Hausdorff et al., 2003; Mbourou, Lajoie, and Teasdale, 2003; Schaafsma et al., 2003) and support the pathological nature of these findings. Stride time and stride length variability exist independent of other factors that can vary across individuals with PD such as gait velocity (Miller, Thaut, McIntosh, and Rice, 1996), average gait cycle duration (Hausdorff et al., 1998; Schaafsma et al., 2003), bradykinesia, rigidity, and tremor (Schaafsma et al., 2003), making measures of the variability in stride time and length good candidates for evaluating motor complications in PD. However, temporal variability of gait may receive greater attention in the literature due to the fact that not all investigators regard the stride length variability a distinct feature of PD gait (Ebersbach, Sojer, et al., 1999; Vieregge et al., 1997).

The increase in variability of the inter-response interval (IRI) during generation of self-paced rhythmic movements (finger tapping) in those with PD may reflect involvement of the BG in temporal control of internally generated rhythmic movements such as gait (Harrington, Haaland, and Hermanowicz, 1998; O'Boyle, Freeman, and Cody, 1996; Pastor, Artieda, Jahanshahi, and Obeso, 1992). According to Wing and colleagues (Wing, Keele, and Margolin, 1984), two mechanisms explain tapping variability. According to their model the central mechanisms issue tap commands at specific time intervals accounting for specific portions of the IRI between consecutive taps. The peripheral component requires a particular average duration to implement this command and account for the remaining portion of the IRI. The strong correlation between the variability of the central component of IRI and the overall IRI variability of finger tapping (Wing et al., 1984), the significant correlation between IRI variability of finger tapping to that of stride duration (Schaafsma et al., 2003), and the occurrence of temporal gait variability without detection of any changes in force amplitude (Baltadjieva et al., 2006; Ebersbach, Heijmenberg, et al., 1999) or without amplitude changes in muscle force (Sheridan and Flowers, 1990) suggest that basal ganglia-cortical pathways may be responsible for increases in temporal gait variability in PD. Significant correlations between the peripheral component of the IRI variability and the variability of IRI of finger tapping and between IRI variability of finger tapping to that of stride duration (Schaafsma et al., 2003) also exist and support a muscular noise influence on temporal gait variability. The fact that muscle noise increases for slower movements (Toxopeus et al., 2011) provides additional support for its influence on temporal gait variability which also increases for slower movements, like those observed in PD. Scientists also blame basal ganglia-cortical pathways for increases in spatial gait variability because pattern of spatial variability for step length and width in mice differ according to the region of BG disrupted by neurotoxins (Amende et al., 2005).

Kinetics. Problems with force production during PD gait initiation persist throughout the gait progression (Dietz, Zijlstra, Prokop, and Berger, 1995). Abnormalities in muscular synergies and loss of functional symmetry of muscle activation complicate force production and lead to gait complications of PD individuals when walking at comfortable speeds (Hausdorff et al., 1998; Miller et al., 1996). Deficits in the BG, resulting in inadequate timing through inadequate and abnormal muscle stimulation, contribute to dysfunctional control of gait patterns. Some scientists suggest an erratic or "noisy" and variable general pattern of muscle activation during gait with a significant reduction in the "rate of force production" (Stelmach, Teasdale, Phillips, and Worringham, 1989). Others show timing of muscle activation similar to healthy older adults (Mitoma et al., 2000) or report impairments only during excessive adaptation to external situations (Dietz et al., 1995), thus for complex walking tasks. Researchers blame (Mitoma et al., 2000) less activation in distal muscle activity, a delay in force production (Stelmach et al., 1989), and over-activity of proximal muscles (Mitoma et al., 2000) for the changes in muscle activity observed during PD gait. Similar to start hesitation, the reduction in activity of the plantar flexors along with insufficient activity of the hip extensor muscles, cause a very weak force production just prior to toe off (Dietz et al., 1995; Mitoma et al., 2000), reducing ground reaction force (Mitoma et al., 2000) and stride length (Vieregge et al., 1997). It appears that both kinematic factors such as decreased range of motion and kinetic factors such as decreased force production contribute to smaller stride length and the subsequent increase in cadence.

The presence of increased variability at early stages of PD, before consumption of any medication, exists (Schaafsma et al., 2003). The close relationship between measures of variability and freezing or falls reflects the significance of these measures in predicting motor complications and possibly in establishing an appropriate evaluation measure for early disease detection or the effectiveness of a treatment.

Freezing of Gait. Freezing of Gait (FOG), a type of akinesia (Narabayashi, 1993), involves a sudden, involuntary, transient, and paroxysmal episode in which the individual becomes incapable of maintaining their current gait pattern and re-starting it, despite arduous efforts (Fahn, 1995; Giladi, Kao, and Fahn, 1997). FOG characterizes a complicated motor symptom in PD with unclear pathophysiology (Giladi et al., 1992) that represents the lack of movement during repetitive and sequential tasks (Narabayashi, 1993). The person produces several tiny steps instead of their regular gait pattern (Ueno et al., 1993) or may stop walking and "freeze in place" (Giladi, Treves, et al., 2001). This complication especially occurs when these people approach a narrow space like a doorway (25% of patients), need to make a turn (45% of patients), cross over an obstacle, shift their attention, or experience additional stress (Giladi et al., 1992; Halliday et al., 1998). This highly debilitating phenomenon, increases the risk of falls, and jeopardizes patients' independence (Gray and Hildebrand, 2000).

The prevalence of FOG ranges from 7% in mild cases to 45% or more in severe cases of PD (Giladi, Treves, et al., 2001; Lamberti et al., 1997). Only 16% of patients under no medication experience this phenomenon (Lamberti et al., 1997), which if present, will be mild and of short duration (Giladi, Treves, et al., 2001). Longer disease duration (>5 years) and longer medication treatment (i.e. levodopa) contribute to prevalence of FOG (Giladi, McDermott, et al., 2001; Giladi et al., 1992; Lamberti et al., 1997). Its prevalence increases in PD persons who demonstrate signs of dementia, dyskinesia, speech disorder, dystonia, postural instabilities, and longer double support time during gait (Giladi et al., 1997; Giladi, McDermott, et al., 2001; Nieuwboer et al., 2004), however for unknown reasons it decreases in those whose disease involves major signs of tremor (Giladi, McDermott, et al., 2001; Lamberti et al., 1997).

FOG often accompanies the hastening phenomenon of festination, where the cadence increases just prior to the freezing (Nieuwboer et al., 2001; Ueno et al., 1993). During this time, shifting the weight between legs takes place more rapidly and appears to be incomplete, as the amplitude of medial to lateral transfer of the center of pressure (COP) decreases (Ueno et al., 1993) and accompanies a decrease in stride length, signifying decreased force production and loss of control over cadence (Nieuwboer et al., 2001). Stride-to-stride duration variability positively correlates to the severity of FOG as indicated by the frequency of FOG episodes (Hausdorff et al., 2003), thus adds to the evidence for loss of temporal control.

In terms of muscular activity, FOG accompanies co-activation and asynchronization of leg muscles (Nieuwboer et al., 2004; Ueno et al., 1993). Unlike the consistent reciprocal pattern of activation in flexor and extensor muscles of the thigh and legs in a healthy individual, co-contraction of flexors and extensors occur in PD patients just prior to freezing (Ueno et al., 1993). The co-contraction of muscles describes a dystonia specific to gait (Giladi, McDermott, et al., 2001). Prominent EMG signals in the tibialis anterior and gastrocnemius muscles of the lower leg appear early and for long durations (Nieuwboer et al., 2004), emphasizing major irregularities in the temporal domain, which also influence spatial irregularity. According to the current literature, freezing stems from *spatial* and *temporal* dysfunctions in PD muscle activation. The greater muscular asynchronization highlights the temporal deficits such as co-contraction possibly leading to inadequate force production and the associated reduction in movement amplitude, like shorter stride length. Limited research in this area makes it difficult to determine causality. However, evidence from other research on motor complications in PD (see section *Gait initiation, Kinetics)* would place emphasis on the basal ganglia-cortical connection deficits.

Some changes in PD gait such as decreases in stride length and gait speed commonly exist in normal older adults (e.g., (Nagasaki et al., 1996)), but do not limit these people's abilities to perform daily tasks such as grocery shopping. Other characteristics such as freezing of gait and shuffling of feet may limit functional performance and/or increase fall risk, and thus present a major concern within this population.

Gait Biomechanics—Gait Termination

Termination of a movement involves the transition from a dynamic state to a static one (Sparrow and Tirosh, 2005). Complications during PD gait termination can create ineffective responses and unsafe conditions.

Kinetics. Complications with gait termination correspond to the PD individuals' general inability to produce sufficient forces to control changes of the displacement of one's COM (Frank, Horak, and Nutt, 2000). To do this the lower limbs must increase the decelerating (braking) forces in the swing limb and decrease the accelerating forces in the stance limb (Bishop, Brunt, Pathare, and Patel, 2002; Hase and Stein, 1998). Unlike healthy individuals, people with PD produce muscle activation patterns opposite to that of controls, leading to inadequate force generation not only during sudden stops (Bishop et al., 2006), but also in conditions requiring a sudden change in direction (Stack, Ashburn, and Jupp, 2006). Abnormal gait termination kinematics compensate for a failure to generate sufficient braking forces to terminate gait.

Kinematics. Gait termination demands control to keep the COM within the BOS at movement end (Pai and Patton, 1997). The inadequate force control in the lower limbs just described leads people with PD to use a compensatory "extra step strategy" to terminate gait and/or reduce their gait velocity earlier in the course of walking to complete the final step with a slower velocity than that of age-matched controls (84% vs 90%) (Bishop et al., 2006).

Clearly, defective force production associated with stopping difficulties in PD must result from ineffective muscle contraction. The limited studies on gait termination in this population make it difficult to offer any conclusions on muscular control. However, one can hypothesize similarities to our previous discussions regarding muscle activation (i.e. inadequate and abnormal muscle stimulation due to BG deficits) exist for gait termination.

Non-Motor Influences on Parkinson's Gait

Non-motor impairments that alter gait kinematics and kinetics occur in many healthy and neurological populations alike also occur in persons with PD. Watch any individual walk while multitasking, under duress, or in the extreme cold and you will likely identify noticeable changes in their gait pattern. Several obvious and subtle alterations in PD gait result from alterations in behavioral, cognitive, sensory, and autonomic symptoms associated with the disease.

Behavioral and Cognitive

Increased cognitive demands, mild cognitive impairment (MCI), dementia, depression, and anxiety influence gait patterns beyond those linked to common PD motor symptoms. Exposure to unpleasant mutilation pictures, which alters the emotional state of participants, reduces step initiation velocity in PD compared to the quickened responses observed in aged-matched controls (Naugle, Hass, Bowers, and Janelle, 2012). Cognitive dual tasking (walking and serial subtraction by 7), an attention demanding task which increases gait variability in PD participants (Hausdorff et al., 2003) similar to older healthy adults (Lundin-Olsson, Nyberg, and Gustafson, 1997), does not correlate with prospective falls in PD (Smulders et al., 2012). However, PD participants with poor attention identified by the Test Battery for Attention Performance also walk slower with a shorter swing phase and a shorter stride than PD participants and controls with better attention (Tseng, Jeng, and Yuan, 2012). PD related dementia (Williams-Gray, Foltynie, Brayne, Robbins, and Barker, 2007) and visuomotor coordination deficits (Uc et al., 2005), linked to cognitive and executive impairments (Inzelberg, Schechtman, and Hocherman, 2008), correlate with characteristic gait and posture motor symptoms of the UPDRS. Non-medicated PD participants with MCI produce a lower single/double leg support time ratio during cognitive dual tasking than participants without PD and MCI and non-medicated PD participants without MCI (Amboni et al., 2012). This ratio during normal gait was also lowest for the PD participants with MCI when both PD groups were medicated. Evidence also supports that executive dysfunction in PD contributes to gait asymmetry (Yogev, Plotnik, Peretz, Giladi, and Hausdorff, 2007) and FOG (Amboni, Cozzolino, Longo, Picillo, and Barone, 2008), the latter of which explains why greater cognitive deficits on the Mini–Mental State Examination or the cognitive section of the Scales for Outcomes in PD reveal higher incidence of FOG (Vercruysse et al., 2012). Moreover, FOG also presents more often in those who report greater anxiety and depression (Giladi and Hausdroff, 2006).

Sensory

Some aspects of vision influence different aspects of PD gait, while others do not. Periodic exposure to the visual environment influences locomotion control in older adults with and without PD similarly (Vitorio et al., 2012). However, increasing or decreasing optic flow patterns (i.e., visual motion cues) relative to those associated with preferred walking speed results in shorter step lengths and larger decreases in walking speed for those with PD when compared to young or elderly control subjects (Schubert, Prokop, Brocke, and Berger, 2005). When combined with reports of dysfunctional dorsal stream projections of visual motion cues in FOG (Lord, Archibald, Mosimann, Burn, and Rochester, 2012), it becomes obvious that visual motion perceptions influence PD gait. In addition, associations between contrast sensitivity and different features of PD gait exist such that those with impaired contrast sensitivity walk slower with shorter steps (Moes and Lombardi, 2009) and report more freezing severity, particularly during locomotion (Davidsdottir, Cronin-Golomb, and Lee, 2005).

Presently, vestibular induced gait abnormalities in PD remain unclear. The benefits of medication used to increase stride length and peak velocity (Blin, Ferrandez, Pailhous, and Serratrice, 1991) during PD gait currently outweigh the associated vestibular deficits observed after administration of medication (Reichert et al., 1982; Weinrich and Bhatia, 1986). These findings explain why vestibular rehabilitation can improve dizziness, balance, and gait outcomes in medicated PD patients (Rossi-Izquierdo, Soto-Varela, Santos-Perez, Sesar-Ignacio, and Labella-Caballero, 2009). The evidence that persons with PD rely more on vestibular inputs compared to proprioceptive inputs for certain task performances (Barnett-Cowan et al., 2010) possibly links to the findings that proprioceptive deficits influence PD motor skills severely (Konczak et al., 2009) and probably more severely than vestibular effects on motor skills.

Sensory system degeneration can contribute to the motor deficits already problematic in individuals with PD. Reductions in plantar sensations, well known to contribute to postural deficits in older adults without PD, also appear in this population (Pratorius, Kimmeskamp, and Milani, 2003) and perhaps contribute to imbalance and gait abnormalities experienced by those with the disease, as improvements occur with plantar surface stimulation during gait (Jenkins et al., 2009). These improvements mimicked those of controls and included increased single-limb support time and decreased peak EMG activation of the tibialis anterior during the foot contact phase of the gait cycle.

Decreased discrimination thresholds in the dorsal toe regions also associate with periodic FOG as medication wears off, thus may contribute to the FOG gait abnormality in this population (Lee, Kim, and Lyoo, 2005). People with PD produce more errors than controls when approaching a target using only proprioceptive feedback (Almeida et al., 2005). Inadequate force matching abilities in PD freezers occur with altered proprioception (Tan, Almeida, and Rahimi, 2011) to help explain existence of these errors. Moreover, many researchers blame this problems on the inadequate sensorimotor integration abilities that accompany basal ganglia deficits (Abbruzzese and Berardelli, 2003).

Autonomic

It appears that gait and autonomic nervous system deficits coexist independently in PD. Researchers associate falling in PD with greater urinary incontinence and longer durations of the Timed-Up-and-Go (TUG) test (Balash et al., 2005). Although these people may try to walk faster to get to the toilet when urinary urges occur, no direct correlation between urinary incontinence and TUG scores currently exist.

The presence of orthostatic hypotension in PD links to increased postural sway, but does not appear to influence speed of gait (Matinolli, Korpelainen, J., Korpelainen, R., Sotaniemi, and Myllyla, 2009). People with PD produce similar oxygen consumption and work load values during cycle ergometer exercise testing to those reported for age- and gender-matched controls (Canning, Alison, Allen, and Groeller, 1997). Moreover, heart rate variability (HRV) does not correlate significantly with gait performance or UPDRS motor scores (Aerts et al.,

2009). HRV does not change during FOG episodes, however heart rate does increase just prior to and during FOG and decrease during sudden stops and turns (Maidan et al., 2010).

Treatments for Parkinson's Gait

Various treatment strategies used to manage motor complications associated with PD exist. Medications relieve certain key symptoms for some but become less effective with advancement of the neural degeneration associated with disease progression. Surgical interventions, currently reserved for qualified individuals who respond less effectively to medication, also relieve certain motor symptoms with fewer side effects than medications. Physical activity can produce benefits for this population that do not occur with pharmacological or surgical treatments. A good treatment program for those with PD incorporates proper exercises, medicine, and in some cases surgical interventions. The following text offers insight to the different intervention strategies used to improve PD motor functions, primarily those associated with dysfunctional gait.

Medications

Pathophysiology of PD directs the medications prescribed for the disease. Administration of medication aims to alleviate motor and cognitive symptoms of PD (Tolosa and Kalzenschlager, 2007). Table 4 summarizes the common drugs administered for relieving motor symptoms of PD, several of which contribute to the different divisions of walking. Medications such as MAO-B inhibitors or Amantadine successfully relieve some of the main motor symptoms in early stages of the disease, however their long term application leads to secondary complications which debilitate PD patients and may cause cognitive problems such as hallucinations (Hely, Morris, Reid, and Trafficante, 2005) and motor dysfunctions such as dyskinesia (Goetz, Poewe, Rascol, and Sampaio, 2005; Hely et al., 2005). With progression of the disease stronger medication (i.e. L-dopa or dopamine agonists) augments or replaces MAO-B inhibitors or Amantadine (Tolosa and Kalzenschlager, 2007). Motor complications such as akinesia, freezing, and postural instabilities (Hely et al., 2005) that contribute to PD gait abnormalities, like changes in stride duration variability (Blin et al., 1991), do not respond well to medications targeting the dopamine system, thus require a different form of treatment. For further information regarding medication for PD see (Goetz et al., 2005; Hely et al., 2005; Tolosa and Kalzenschlager, 2007).

Surgery

Surgical interventions used for symptomatic treatments do not lead to the debilitating side effects present with medication usage (Thobois, Delamarre-Damier, and Derkinderen, 2005). Those with PD who do qualify for surgical procedures should consult with their movement disorder neurologist and neurosurgeon to ensure the life changing benefits compensate for the surgical risks before agreeing to such an invasive procedure.

Table 4. Common anti-Parkinson's medications, characteristics, advantages and disadvantages

Drug	Characteristics	Advantages	Disadvantages
L-dopa	-Dopamine precursor (often administered with catechol-O-methyltransferase (COMT) inhibitors (entacapone, tolacapone) to prolong half life of L-dopa.)	-Relieves major motor symptoms -Increases life expectancy -Improves quality of life -Most effective, *gold standard* drug	-Causes motor fluctuations -Causes dyskinesia -Does not affect freezing, dysautonomia, dysarthria -Causes hallucinations, nausea, sleepiness, autonomic problems
Dopamine agonists bromocriptines, cabergoline, di-hydroergocryptine, lisuride, pergolide, pribedil, pram-ipexole, ropinirole	-Stimulate dopamine receptors	-Fewer motor fluctuations and complications compared to L-dopa -Antioxidant effect -Neural protective	-Causes nausea, vomiting, bradycardia, postural hypotension, dizziness, sleepiness, hallucination, psychosis, insomnia, dyskinesia -Beneficial for 3-5 years
MAO-B inhibitors: Sekegukube (deprenyl), rasagiline	-Increases half life of dopamine in the brain by inhibiting oxidative deamination of dopamine	-Delays start of L-dopa usage -Reduces symptom severity -Reduces motor fluctuations	-Insomnia -Causes hallucinations, nausea, dyskinesia, confusion, autonomic side effects (orthostatic hypotension) with L-dopa
Amantadine	-Enhances release of dopamine from dopamine terminals. -Increases affinity of dopamine receptors -Exact mechanism unknown	-Delays start of L-dopa usage -Reduces akinesia, tremor, rigidity -Mild side effects compared to L-dopa	-Efficacy reduces over time -Causes livedo reticularis, ankle edema, mouth dryness, blurred vision, difficulty focusing, confusion, depression, nightmares, hallucinations

Adopted from (Tolosa and Kalzenschlager, 2007).

Surgical interventions for PD may involve severing certain parts of the basal ganglia as with pallidotomy, thalamotomy, and subthalamotomy (Espay, Mandybur, and Revilla, 2006) or less abrasive methods such as deep brain stimulation (DBS) (Bhidayasiri, Srikijvilaikul, Locharernkul, Phanthumchinda, and Kaoroptham, 2006; Espay et al., 2006; Esselink et al., 2006; Pahwa et al., 2001). Preference for the latter, more common procedure, results from similar benefits with fewer side effects. Advantages of DBS over neural removal include decreased demand for re-operation, lower morbidity rates, reversibility, larger improvements in motor and non-motor symptoms (Bhidayasiri et al., 2006; Esselink et al., 2006), and decreased dose demands for medication (Benabid et al., 2000; Kumar et al., 1998).

DBS can target different nuclei within the BG and its connections. Targeting the thalamus eliminates signs of tremor (Koller et al., 1997) without affecting other motor symptoms (Limousin, Speelman, Gielen, and Janssens, 1999). Stimulation of GPi reduces signs of drug induced dyskinesia and major PD motor symptoms such as bradykinesia, akinesia, and rigidity (Dostrovsky, Hutchison, and Lozano, 2002), but the effectiveness declines after the first year of the surgery (Volkmann et al., 2004). The most common procedure of STN stimulation produces general improvement in motor function (58%) and reductions in tremor (82%), akinesia (57%), rigidity (52%), postural instability (49%), and drug-induced dyskinesia (83%) (Kumar et al., 1998).

Investigators suggest that DBS works by regulating the neural oscillation of the targeted nuclei (Welter et al., 2004). Stimulation appears to terminate the dysfunctional activity of neurons and shifts the temporal pattern of their discharge to a more optimal and synchronized frequency (Hashimoto, Elder, Okun, Patrick, and Vitek, 2003). This pattern of activity replaces the random deleterious pattern of BG network with a regular pattern of neural bursts causing a stabilized and symmetric activity across different regions within the BG (McIntyre, Savasta, Kerkerian-Le Goff, and Vitek, 2004).

Improvements in motor symptoms with DBS of the STN do not necessarily indicate improvements in aspects of PD gait. Enhancements in gait observed with stimulation include increases in stride length, gait velocity, hip and knee range of motion (Piper, Abrams, and Marks, 2005), and improvements in FOG (Niu et al., 2012). Other reports also include no changes in functional gait or mobility (Kelly et al., 2010), walking cadence, and ankle excursions when walking (Piper et al., 2005). Greater duration preparing to initiate gait with smaller first steps (Rocchi et al., 2012) suggest stimulation can worsen step initiation characteristics. While stimulation of the dorsal STN produces improvements in step length and velocity not observed with ventral STN stimulation (Johnsen, Sunde, Mogensen, and Ostergaard, 2010), precise stimulation area can explain some of the different outcomes observed with DBS. Clearly, gait related improvements do not always accompany DBS and require further investigation.

Benefits of DBS, although promising, often remain most effective when used in conjunction with patients' regular medication (Goetz et al., 2005). Unfortunately, not all PD patients can undergo such surgery. The suitable surgery candidate responds to medication and is less than 75 years of age with relatively normal cognitive status and disease onset of greater than 5 years (Amick and Grace, 2006; Chang and Chou, 2006; Rodriguez, Fernandez, Haq, and Okun, 2007). Therefore, research continues for discovering more effective medical and surgical techniques, and includes less invasive techniques such as exercise.

Exercise Training

A properly designed physical activity program will help individuals with PD enhance or overcome disease-related motor symptoms needed to improve physical function. The ability to generate correct movements remains intact in PD (Robertson and Flowers, 1990), however the major motor complications rise from the loss of an internal ability to activate the appropriate motor plan or suppress the unwanted plans (Nambu, 2005; Robertson and Flowers, 1990). Different rehabilitation techniques through exercise training or therapy should focus on activating the correct motor plan. To do so effectively one should understand the disease, individual characteristics of the patient, medications used with peak effect time (on/off periods), non-motor impairments, and presence of other pathologies (Morris, 2000).

Many diverse exercise protocols reveal significant functional benefits for those with PD. Not all exercise strategies produce the same benefits in individuals with and without PD, however several motor learning strategies produce motor improvements, including those associated with the different divisions of gait.

Table 5 lists some guidelines to help develop successful rehabilitation protocols in PD. The first of these guidelines resembles early stages of progressive training protocols, deemed beneficial for PD training (David et al., 2012; Hondzinski, 2012). Progressive training protocols also utilize recurring practice and advances in movement difficulty over time from the motor learning literature. These approaches promote positive performance outcomes in the initial training stages to encourage adherence (Bandura, 1997) and continue to promote greater positive performance outcomes later in training due to advances in exercise intensity (Farley and Koshland, 2005; Ridgel, Vitek, and Alberts, 2009). Escalation in exercise intensity appears to produce some of the most successful long term motor outcomes in PD (e.g., decreases in disease severity (Corcos et al., 2012)). The second guideline makes use of cortico-striatal pathways not directly linked to PD neural degeneration. Bringing attention to a particular task through instructions can reveal at least short term improvements in PD gait. For example, walking velocity and step length in people with PD increased with instructions emphasizing either arm swinging or large stepping (Behrman, Teitelbaum, and Cauraugh, 1998). The third major guideline for successful training in this population involves visual, auditory, or tactile cueing. External cueing signals may bypass the damaged basal ganglia and make use of intact neural structures and pathways in this populace. PD participants who trained with rhythmic auditory cues revealed functional gait improvements which lasted longer after training termination than PD participants who trained without such cueing (Kadivar, Corcos, Foto, and Hondzinski, 2011). Different improvements in gait kinetics and kinematics with different training techniques provide insight to successful exercise protocols for PD gait.

PD participants reveal greater improvements in motor function and gait biomechanics with greater training intensity in different exercise domains. Eighteen to twenty sessions of twice weekly progressive lower limb resistance training resulted in increased COP posterior displacement, stride length, and stride velocity of gait initiation for PD participants (Hass, Buckley, Pitsikoulis, and Barthelemy, 2012). As expected with healthy subjects, these pre-post training improvements exceeded those of PD controls who did not change their daily activity patterns during the training period. PD participants who performed stretching, walking, cycling, and high intensity progressive resistance exercises three days a week for 12 weeks produced greater increases in quadriceps muscle volume, distance on the six minute walk test, decreases in stair descent time (Dibble et al., 2006), force production in the quadriceps muscles, increases in gait speed, and decreases in TUG duration (Dibble, Hale, Marcus, Gerber, and LaStayo, 2009) relative to participants with PD who trained similarly but with normal intensity resistance exercises. Individuals with PD participating in either (1) 5 no intensity PD education classes, (2) 24 progressive high intensity treadmill sessions plus education, or (3) 24 low intensity physical therapy sessions (stretching, range of motion, balance, resistance training, functional, and transitional tasks) plus education over 8 weeks revealed improvements in gait velocity (Fisher et al., 2008). Interestingly, post training increases in step length and excursions of hip and ankle joints accompanied increases in self-selected and fast gait speed selections to the greatest extent in the high intensity group. Moreover, ground reaction forces during the sit-to-stand transition increased in symmetry after training completion only for this group. Clearly, individuals with PD tolerate various progressive high intensity exercises well while reaping the associated benefits.

Table 5. Guidelines for developing rehabilitation strategies

1	Avoid complexity early in training; break movement/movement sequence into components; teach only one task at a time
2	Cortically control movement; use attention to compensate for the loss automaticity
3	Use external cues; visual, auditory, and tactile cues to start, pace, or end movement

Adapted from (Iansek, 1999; Morris, 2000).

Good high intensity training programs like those just mentioned include aspects of training progression including relative intensity increases that allow one to increase speed, resistance, or time of exercise as they improve. However, improvements in gait biomechanics can occur under other conditions too.

Task-specific training works very well in improving PD gait. Gait initiation exercises involving repetitive lateral hip pulls during step initiation shortened step duration, thus improved step initiation (Mille et al., 2009). Dual tasking involving word recall, serial subtraction by 3, and simple math while walking three times per week for four weeks resulted in gait speed increases and stride time variability decreases with and without cognitive loads (Yogev-Seligman, Giladi, Brozgol, and Hausdorff, 2012). Six weeks of treadmill walking while avoiding virtual obstacles produced faster gait speeds and longer step lengths in PD participants who walked over ground with and without maneuvers through obstacles (Mirelman, Maidan, Deutsch, Giladi, and Hausdorff, 2011). Note that task-specific training in each case resulted in improvements in the division of gait trained, whether context remained the same or not.

PD gait improvements occur with training protocols that incorporate concurrent sensory cueing strategies. Single and/or multiple session training paradigms involving tactile, visual, auditory, or multiple sensory cueing produce gait improvements in people with PD.

Although less common than the use of auditory or visual cues, PD gait changes can occur with use of tactile sensory cueing during training. Several walking trials with stimulating insoles, which increase plantar sensory stimulation, caused increased duration in single-limb support time and earlier peak activation of the tibialis anterior in people with PD (Jenkins et al., 2009). When compared to walking with normal insoles, vibratory insoles below the heel and forefoot which provided stimulation during stance, increased stride duration and length, gait speed, and cadence and decreased stride variability in individuals with PD (Novak, P. and Novak, V., 2006). These responses to added plantar tactile stimulation approached or mimicked those of control subjects, thus appeared to normalize certain aspects of gait. The specificity of and accessibility to equipment for such stimulation may explain the relatively rare use of tactile cueing as a training protocol.

The more common use of rhythmic auditory stimulation (RAS) in training can produce spatial and temporal modifications in PD gait deemed as improvements. Researchers, clinicians, and trainees can access RAS through use of a metronome or embedded in music and increase or decrease RAS frequency to vary training intensity. Concurrent use of RAS with walking produces several gait improvements including reductions in step variability during turns (Willems et al., 2007) and declines in the number of FOG episodes (Arias and

Cudeiro, 2010), two major difficulties associated with PD gait. Single training sessions using RAS can elicit increases in gait velocity, step length (McIntosh, Brown, Rice, and Thaut, 1997), and cadence and produce decreases in double support time (Freedland et al., 2002). Weeks of gait training with concurrent RAS use increased gait velocity, stride length (Thaut et al., 1996), and cadence (del Olmo and Cudeiro, 2005), reduced variability and asymmetry of tibialis anterior and gastrocnemius muscles (Miller et al., 1996), but did not alter the number of freezing episodes (Cubo, Leurgans, and Goetz, 2004) during non-cued gait performances compared to pre-training values. In fact, adding RAS to six weeks of progressive multidirectional step training not only produced improvements in PD functional gait as indicated by the Dynamic Gait Index, TUG, and Tinetti-gait scores, it also allowed these individuals to maintain improvements at least eight weeks after training termination and longer than those trained without RAS (Kadivar et al., 2011). Moreover, PD participants with cognitive impairment increased walking speed and step frequency of single and dual task gait after practice including RAS and attention of focus (Rochester, Burn, Woods, Godwin, and Nieuwboer, 2009). Tasks involved walking with and without carrying a tray with two cups of water. Practice in each task involved three trials without cues, two trials with RAS and attention emphasis on stepping with the RAS beat, and two trials with RAS and emphasis on taking big steps. These varied successes along with easy access and the adaptable capacity of RAS use help explain its popularity in training for this population.

The popularity of exercise training with visual cues in PD likely stems from the idea that this population possesses an over reliance on vision (Schubert et al., 2005). Increases in stride length and gait velocity when walking straight and decreases in completion time, number of steps, and number of freezing episodes during turns existed for conditions in which PD participants walked with a cane projecting a transverse green light across the walking path, relative to performances with a cane projecting a red light or no light (Bryant, Rintala, Lai, and Protas,, 2010). Clearly one should be aware of color choice, which may be linked to contrast sensitivity, for visual cueing use during gait. Concurrent use of transverse white visual cue strips while walking over ground increased gait velocity, stride length, and knee angle excursion, and decreased relative stance duration (de Melo Roiz, Azevedo Cacho, Cliquet, and Barasnevicius Quaqliato, 2011). Similar simultaneous visual cueing during externally paced treadmill walking resulted in increased stride length and stride time, decreased cadence, and reduced stride time variability at a low velocity of 1 km/hr (Luessi, Mueller, Breimhorst, and Vogt, 2012). The similar improvements in stride length between the two conditions, regardless of differences between the over ground walking and treadmill walking, likely resulted from the forced gait speed in the latter. Transverse lines provided during treadmill or over ground training revealed increased step length and walking velocity during non-cued over ground walking which were maintained 6 weeks after training termination (Almeida and Bhatt, 2012). Over ground training with visual cues also revealed functional gait improvements, assessed by the TUG, and maintenance of these improvements to suggest that improvements in certain aspects of gait do not require motion cues provided during over ground walking. Interestingly, artificially induced relative visual motion cues signifying faster movements than those associated with actual walking speed induced increases in gait velocity and step length (Schubert et al., 2005). Relative visual motion cues signifying slower movements induced the corresponding decreases in gait velocity and step length. These results show that different types of visual cues can alter PD gait.

Adaptations in PD gait biomechanics also occur after training which involves multiple cueing strategies. Auditory and visual cues provided during 4 weeks of progressive treadmill training and progressive over ground gait training resulted in improvements in reported FOG, longer distances in the 6MW, and increased gait speed and stride length (Frazzitta, Maestri, Uccellini, Bertotti, and Abelli, 2009). The forced increases in gait speed with treadmill training likely explains the greater improvements observed for those trained on the treadmill compared to those trained over ground, as mentioned previously forcing greater training intensity can produce greater movement outcomes in people with PD (Fisher et al., 2008).

Three weeks of training gait initiation, termination, dual tasking, and forward, backward, and turning progression on various surfaces and through tight spaces with auditory, visual, and tactile cueing for the first week and a preferred cueing stimulus thereafter resulted in non-cued decreases in freezing frequency and increases in gait speed, step length (Nieuwboer et al., 2009), and walking time (Lim et al., 2010) relative to pre-training values. Moreover, increases in walking speed and step length during non-cued single and dual task walking after training lasted for at least 6 weeks (Rochester et al., 2010). Obviously, multi-cue training protocols can produce gait alterations in people with PD that could last through periods off training.

Various training strategies which challenge individuals with PD can succeed in targeting specific gait limitations caused by the disease. Value of each intervention may vary by individual according to disability, convenience, and cost-effectiveness. People with PD should continue or resume physical exercise as part of their treatment for the disease, as regular training may not only modify gait dysfunction, but also disease progression (Ransmayr, 2011).

Conclusion

Motor, behavioral, cognitive, sensory, and/or autonomic system declines associated with PD contribute to abnormal gait kinetics and kinematics. However, the heterogeneity of disability across this population and through disease advancement demands the use of individualized treatment programs which incorporate proper medication and exercise, and in some cases, the use of surgery. Together, these treatments can and should improve not only the mobility in those with PD but also the quality of life.

References

Aarsland, D. Bronnick, K., Fladby, T. (2011). Mild cognitive impairment in Parkinson's disease. *Current Neurology and Neuroscience Reports,* 11, 371-378. doi: 10.1007/s11910-011-0203-1

Abbruzzese, G. and Berardelli, A. (2003). Sensorimotor integration in movement disorders. *Movement Disorders,* 18(3), 231-240.

Adamovich, S. V., Berkinblit, M. B., Hening, W., Sage, J., and Poizner, H. (2001). The interaction of visual and proprioceptive inputs in pointing to actual and remembered targets in Parkinson's disease. *Neuroscience,* 104(4), 1027-1041.

Aerts, M., Synhaeve, N., Mirelman, A., Bloem, B., Giladi, N., and Hausdorff, J. (2009). Is heart rate variability related to gait impairment in patients with Parkinson's disease? A pilot study. *Parkinsonism and Related Disorders,* 15, 712-715.

Albin, R. L., Young, A. B. and Penney, J. B. (1995). The functional anatomy of disorders of the basal ganglia. *Trends in Neurosciences,* 18(2), 63-64.Alexander, G. E., DeLong, M. R. and Strick, P. L. (1986). Parallel organization of functionally segregated circuits linking basal ganglia and cortex. *Annual Review of Neuroscience,* 9, 357-381.

Allcock, L., Ullyart, K., Kenny, R., and Burn, D. (2004). Frequency of orthostatic hypotension in a commnity based cohort of patients with Parkinson's disease. *Journal of Neurology, Neurosurgery, and Psychiatry,* 75, 1470-1471. doi: 10.1136/jnnp.2003.029413

Almeida, Q., Bhatt, H. (2012). A manipulation of visual feedback during gait training in Parkinson's disease. *Parkinson's Disease*. doi: 10.1155/2012/508720

Almeida, Q. J., Frank, J. S., Roy, E. A., Jenkins, M. E., Spaulding, S., Patla, A. E., and Jog, M. S. (2005). An evaluation of sensorimotor integration during locomotion toward a target in Parkinson's disease. *Neuroscience,* 134(1), 283-293.

Amboni, M., Cozzolino, A., Longo, K., Picillo, M., and Barone, P. (2008). Freezing of gait and executive functions in patients with Parkinson's disease. *Movement Disorders,* 23(3), 395-400.

Amboni, M., Barone, P., Iuppariello, L., Lista, I., Tranfaglia, R., Fasano, A., … Sorrentino, G. (2012). Gait patterns in parkinsonian patients with or without mild cognitive impairment. *Movement Disorders,* 27(12), 1536-1543. doi: DOI: 10.1002/mds.25165

Amende, I., Kale, A., McCue, S., Glazier, S., Morgan, J. P., and Hampton, T. G. (2005). Gait dynamics in mouse models of Parkinson's disease and Huntington's disease. *Journal of NeuroEngineering* and *Rehabilitation,* 2, 20.

Amick, M. M. and Grace, J. (2006). Deep brain stimulation surgery for Parkinson's disease: the role of neuropsychological assessment. *Medicine and Health, Rhode Island,* 89(4), 130-133.

Anderson, G., Noorian, A. R., Taylor, G., Anitha, M., Bernhard, D., Srinivasan, S., and Greene, J. G. (2007). Loss of enteric dopaminergic neurons and associated changes in colon motility in an MPTP mouse model of Parkinson's disease. *Experimental Neurology,* 207(1), 4-12. doi: http://dx.doi.org/10.1016/j.expneurol.2007.05.010

Arias, P. and Cudeiro, J. (2010). Effect of rhythmic auditory stimulation on gait in Parkinsonian patients with and without freezing of gait. *PLoS ONE,* 5(3), e9675. doi: 10.1371/journal.pone.0009675

Armstrong, R. (2011). Visual symptoms in Parkinson's disease. *Parkinson's Disease*. doi: 10.4061/2011/908306

Balash, Y., Peretz, C., Leibovich, G., Herman, T., Hausdorff, J. M., and Giladi, N. (2005). Falls in outpatients with Parkinson's disease: frequency, impact and identifying factors. *Journal of Neurology,* 252(11), 1310-1315.

Baltadjieva, R., Giladi, N., Gruendlinger, L., Peretz, C., and Hausdorff, J. M. (2006). Marked alterations in the gait timing and rhythmicity of patients with de novo Parkinson's disease. *European Journal* of *Neuroscience,* 24(6), 1815-1820.

Bandura, A. (1997). *Self efficacy: the exercise of control.* New York: W. H. Freeman and Company.

Barnett-Cowan, M., Dyde, R., Fox, S., Moro, E., Hutchison, W., and Harris, L. (2010). Multisensory determinants of orientation perception in Parkinson's disease. *Neuroscience,* 167, 1138-1150.

Behrman, A. L., Teitelbaum, P. and Cauraugh, J. H. (1998). Verbal instructional sets to normalise the temporal and spatial gait variables in Parkinson's disease. *Journal of Neurology, Neurosurgery, and Psychiatry,* 65(4), 580-582.

Benabid, A. L., Krack, P. P., Benazzouz, A., Limousin, P., Koudsie, A., and Pollak, P. (2000). Deep brain stimulation of the subthalamic nucleus for Parkinson's disease: methodologic aspects and clinical criteria. *Neurology,* 55(12 Suppl. 6), S40-44.

Bhidayasiri, R., Srikijvilaikul, T., Locharernkul, C., Phanthumchinda, K., and Kaoroptham, S. (2006). State of the art: deep brain stimulation in Parkinson's disease. *Journal of the Medical Association* of *Thailand,* 89(3), 390-400.

Bishop, M., Brunt, D. and Marjama-Lyons, J. (2006). Do people with Parkinson's disease change strategy during unplanned gait termination? *Neuroscience Letters,* 397(3), 240-244.

Bishop, M. D., Brunt, D., Pathare, N., and Patel, B. (2002). The interaction between leading and trailing limbs during stopping in humans. *Neuroscience Letters,* 323(1), 1-4.

Blin, O., Ferrandez, A. M., Pailhous, J., and Serratrice, G. (1991). Dopa-sensitive and dopa-resistant gait parameters in Parkinson's disease. *Journal of the Neurological Sciences,* 103(1), 51-54.

Blin, O., Ferrandez, A. M. and Serratrice, G. (1990). Quantitative analysis of gait in Parkinson patients: increased variability of stride length. *Journal of the Neurological Sciences,* 98(1), 91-97.

Bloem, B. R., Grimbergen, Y. A., Cramer, M., Willemsen, M., and Zwinderman, A. H. (2001). Prospective assessment of falls in Parkinson's disease. *Journal of Neurology,* 248(11), 950-958.

Bloem, B. R., Hausdorff, J. M., Visser, J. E., and Giladi, N. (2004). Falls and freezing of gait in Parkinson's disease: a review of two interconnected, episodic phenomena. *Movement Disorders,* 19(8), 871-884. doi: 10.1002/mds.20115

Bronstein, A. M., Brandt, T., Woolacott, M. H., and Nutt, J. G. (2004). *Clinical Disorders of Balance, Posture And Gait* (2 ed.). Euston Road: Arnold.

Bryant, M., Rintala, D., Lai, E., and Protas, E. (2010). A pilot study: influence of visual cue color on freezing of gait in persons with Parkinsons disease. *Disability and Rehabilitation. Assistive Technology,* 5(6), 456-461. doi: 10.3109/17483107.2010.495815

Canning, C. G., Alison, J. A., Allen, N. E., and Groeller, H. (1997). Parkinson's disease: an investigation of exercise capacity, respiratory function, and gait. [Research Support, Non-US Gov't]. *Archives of Physical Medicine* and *Rehabilitation, 78*(2), 199-207.

Chang, V. C. and Chou, K. L. (2006). Deep brain stimulation for Parkinson's disease: patient selection and motor outcomes. *Medicine and Health, Rhode Island,* 89(4), 142-144.

Corcos, D. M., Robichaud, J. A., David, F. J., Leurgans, S. E., Vaillancourt, D. E., Poon, C., ... Comella, C. L. (2013). A two-year randomized controlled trial of progressive resistance exercise for Parkinson's disease. *Movement Disorders, 28*(9), 1230-1240. doi: 10.1002/mds.25380

Crawford, T., Goodrich, S., Henderson, L., Kennard, C. (1989). Predictive responses in Parkinson's disease: manual keypresses and saccadic eye movements to regular stimulus events. *Journal of Neurology, Neurosurgery, and Psychiatry,* 52, 1033-1042. doi: 10.1136/jnnp.52.9.1033

Crenna, P. and Frigo, C. (1991). A motor programme for the initiation of forward-oriented movements in humans. The Journal of *Physiology,* 437, 635-653.

Cubo, E., Leurgans, S. and Goetz, C. G. (2004). Short-term and practice effects of metronome pacing in Parkinson's disease patients with gait freezing while in the 'on' state: randomized single blind evaluation. *Parkinsonism and Related Disorders,* 10(8), 507-510.

Dastgheib, Z., Lithgow, B. and Moussavi, Z. (2012). Diagnosis of Parkinson's disease using electrovestibulography. *Medical and Biological Engineering and Computing,* 50, 483-491. doi: 10.1007/s11517-012-0890-z

David, F., Rafferty, M., Robichaud, J., Prodoehl, J., Kohrt, W., Vaillancourt, D., and Corcos, D. (2012). Progressive resistance exercise and Parkinson's disease: a review of potential mechanisms. *Parkinson's Disease,* 2012. doi: 10.1155/2012/124527

Davidsdottir, S., Cronin-Golomb, A. and Lee, A. (2005). Visual and spatial symptoms in Parkinson's disease. *Vision Research,* 45, 1285-1296.

De Melo Roiz, R., Azevedo Cacho, E., Cliquet Jr., A., and Barasnevicius Quagliato, E. (2011). Analysis of parallel and transverse visual cues on the gait of individuals with idiopathic Parkinson's disease. *International Journal of Rehabilitation Research,* 34(4), 343-348. doi: 10.1097/MRR. 0b013e32834d32f0

Del Olmo, M. F. and Cudeiro, J. (2005). Temporal variability of gait in Parkinson disease: effects of a rehabilitation programme based on rhythmic sound cues. *Parkinsonism and Related Disorders,* 11(1), 25-33.

DeLong, M. R. (1990). Primate models of movement disorders of basal ganglia origin. *Trends in Neurosciences,* 13(7), 281-285.

Deuschl, G., Volkmann, J. and Raethjen, J. (2007). Tremors: Differential diagnosis, pathophysiology and therapy. In: J. Jankovic and T. Eduardo (Eds.), *Parkinson's Disease and Movement Disorders* (5 ed., pp. 306-308). Philadelphia: Lippincott Williams and Wilkins.

Dibble, L., Hale, T., Marcus, R., Droge, J., Gerber, J., and LaStayo, P. (2006). High-intensity resistance training amplifies muscle hypertrophy and functional gains in persons with Parkinson's disease. *Movement Disorders,* 21(9), 1444-1452.

Dibble, L., Hale, T., Marcus, R., Gerber, J., and LaStayo, P. (2009). High intensity eccentric resistance training decreases bradykinesia and improves quality of life in persons with Parkinson's disease: a preliminary study. *Parkinsonism and Related Disorders,* 15, 752-757.

Dietz, V., Leenders, K. L. and Colombo, G. (1997). Leg muscle activation during gait in Parkinson's disease: influence of body unloading. *Electroencephalography and Clinical Neurophysiology,* 105(5), 400-405.

Dietz, V., Zijlstra, W., Prokop, T., and Berger, W. (1995). Leg muscle activation during gait in Parkinson's disease: adaptation and interlimb coordination. *Electroencephalography and Clinical Neurophysiology,* 97(6), 408-415.

Donoghue, J. P. and Herkenham, M. (1986). Neostriatal projections from individual cortical fields conform to histochemically distinct striatal compartments in the rat. *Brain Research,* 365(2), 397-403.

Dostrovsky, J. O., Hutchison, W. D. and Lozano, A. M. (2002). The globus pallidus, deep brain stimulation, and Parkinson's disease. *Neuroscientist,* 8(3), 284-290.

Ebersbach, G., Heijmenberg, M., Kindermann, L., Trottenberg, T., Wissel, J., and Poewe, W. (1999). Interference of rhythmic constraint on gait in healthy subjects and patients with early Parkinson's disease: evidence for impaired locomotor pattern generation in early Parkinson's disease. *Movement Disorders,* 14(4), 619-625.

Ebersbach, G., Sojer, M., Valldeoriola, F., Wissel, J., Muller, J., Tolosa, E., and Poewe, W. (1999). Comparative analysis of gait in Parkinson's disease, cerebellar ataxia and subcortical arteriosclerotic encephalopathy. *Brain,* 122 (Pt 7), 1349-1355.

Espay, A. J., Mandybur, G. T. and Revilla, F. J. (2006). Surgical treatment of movement disorders. *Clinics in Geriatric Medicine,* 22(4), 813-825, vi.

Esselink, R. A., de Bie, R. M., de Haan, R. J., Steur, E. N., Beute, G. N., Portman, A. T., ... Speelman, J. D. (2006). Unilateral pallidotomy versus bilateral subthalamic nucleus stimulation in Parkinson's disease: one year follow-up of a randomised observer-blind multi centre trial. *Acta Neurochirurgica,* 148(12), 1247-1255.

Fahn, S. (1995). The freezing phenomenon in parkinsonism. *Advances in Neurology,* 67, 53-63.

Fahn, S., Elton, R. L. and Members of the UPDRS Development Committee. (1987). In: Fahn, S., Marsden, C.D., Calne D.B., Goldstein M., eds. *Recent developments in Parkinson's disease* (Vol. 2). Florham Park, NJ: Macmillan Healthcare Information.

Farley, B., Koshland, G. (2005). Training BIG to move faster: the application of the speed-amplitude relation as a rehabilitation strategy for people with Parkinson's disease. *Experimental Brain Research,* 167, 462-467.

Fisher, B., Wu, A., Salem, G., Song, J., Lin, C., Yip, J., Cen, S., Gordon, J., Jakowec, M., and Petzinger, G. (2008). The effect of exercise training in improving motor performance and corticomotor excitability in persons with early Parkinson's disease. *Archives of Physical Medicine and Rehabilitation,* 89(7), 1221-1229. doi: 10.1016/j.apmr.2008.01.013

Frank, J. S., Horak, F. B. and Nutt, J. (2000). Centrally initiated postural adjustments in parkinsonian patients on and off levodopa. *Journal of Neurophysiology,* 84(5), 2440-2448.

Frazzitta, G., Maestri, R., Uccellini, D., Bertotti, G., and Abelli, P. (2009). Rehabilitation treatment of gait in patients with Parkinson's disease with freezing: a comparison between two physical therapy protocols using visual and auditory cues with or without treadmill training. *Movement Disorders,* 24(8), 1139-1143.

Freedland, R. L., Festa, C., Sealy, M., McBean, A., Elghazaly, P., Capan, A., ... Rothman, J. (2002). The effects of pulsed auditory stimulation on various gait measurements in persons with Parkinson's Disease. *NeuroRehabilitation,* 17(1), 81-87.

Fung, V. S. C. and Thompson, P. D. (2007). Rigidity and Spasticity. In: J. Jankovic and T. Eduardo (Eds.), *Parkinson's Disease and Movement Disorders* (5 ed., pp. 504-513). Philadelphia: Lippincott Williams and Wilkins.

Gabell, A. and Nayak, U. S. (1984). The effect of age on variability in gait. *J. Gerontol.,* 39 (6), 662-666.

Gancher, S. T. (2006). Parkinson's Disease. In: R. M. Herndon (Ed.), *Handbook of Neurological Rating Scales* (2 ed., pp. 145-154). New York: Demos Medical Publishing.

Gantchev, N., Viallet, F., Aurenty, R., and Massion, J. (1996). Impairment of posturo-kinetic co-ordination during initiation of forward oriented stepping movements in parkinsonian patients. *Electroencephalography and Clinical Neurophysiology,* 101(2), 110-120.

Garcia-Rill, E. (1991). The pedunculopontine nucleus. *Progress in Neurobiology,* 36(5), 363-389.

Georgiou, N., Iansek, R., Bradshaw, J. L., Phillips, J. G., Mattingley, J. B., and Bradshaw, J. A. (1993). An evaluation of the role of internal cues in the pathogenesis of parkinsonian hypokinesia. *Brain,* 116 (Pt 6), 1575-1587.

Giladi, N. and Hausdroff, J. (2006). The role of mental function in the pathogenesis of freezing of gait in Parkinson's disease. *Journal of the Neurological Sciences,* 248, 173-176.

Giladi, N., Balash, Y., Ruzicka, E., and Jankovic, J. (2007). Disorders of gait. In: J. Jankovic and T. Eduardo (Eds.), *Parkinson's Disease and Movement Disorders* (5 ed., pp. 446). Philadelphia Lippincott Williams and Wilkins.

Giladi, N., Kao, R. and Fahn, S. (1997). Freezing phenomenon in patients with parkinsonian syndromes. *Movement Disorders,* 12(3), 302-305.

Giladi, N., McDermott, M. P., Fahn, S., Przedborski, S., Jankovic, J., Stern, M., and Tanner, C. (2001). Freezing of gait in PD: prospective assessment in the DATATOP cohort. *Neurology,* 56(12), 1712-1721.

Giladi, N., McMahon, D., Przedborski, S., Flaster, E., Guillory, S., Kostic, V., and Fahn, S. (1992). Motor blocks in Parkinson's disease. *Neurology,* 42(2), 333-339.

Giladi, N., Treves, T. A., Simon, E. S., Shabtai, H., Orlov, Y., Kandinov, B., ... Korczyn, A. D. (2001). Freezing of gait in patients with advanced Parkinson's disease. *Journal of Neural Transmission,* 108(1), 53-61.

Glass, G. A., Josephs, K. A. and Ahlskog, J. E. (2006). Respiratory insufficiency as the primary presenting symptom of multiple-system atrophy. *Archives of Neurology,* 63, 978-981.

Goetz, C., Poewe, W., Rascol, O., Sampaio, C., Stebbins, G. (2003). The unified Parkinson's disease rating scale (UPDRS): status and recommendations. *Movement Disorders,* 18(7), 738-750.

Goetz, C., Poewe, W., Rascol, O., Sampaio, C., Stebbins, G., Counsell, C., ... Seidl, L. (2004). Movement Disorder Society Task Force report on the Hoehn and Yahr staging scale: status and recommendations. *Movement Disorders,* 19(9), 1020-1028.

Goetz, C. G., Poewe, W., Rascol, O., and Sampaio, C. (2005). Evidence-based medical review update: pharmacological and surgical treatments of Parkinson's disease: 2001 to 2004. *Movement Disorders,* 20(5), 523-539.

Goetz, C. G., Tilley, B. C., Shaftman, S. R., Stebbins, G. T., Fahn, S., Martinez-Martin, P., ... LaPelle, N. (2008). Movement disorder society-sponsored revision of the unified parkinson's disease rating scale (MDS-UPDRS): scale presentation and clinimetric testing results. *Movement Disorders,* 23(15), 2129-2170.

Goldstein, D. (2006). Cardiovascular aspects of Parkinson disease. *Journal of Neural Transmission Supplementa,* 70, 339-342.

Gorell, J. M., Ordidge, R. J., Brown, G. G., Deniau, J. C., Burderer, N. M., and Helpern, J. A. (1995). Increased iron-related MRI contrast in the substantia nigra in Parkinson's disease. *Neurology,* 45(6), 1138-1143.

Gray, P. and Hildebrand, K. (2000). Fall risk factors in Parkinson's disease. *Journal of Neuroscience* Nursing, 32(4), 222-228.

Halliday, S. E., Winter, D. A., Frank, J. S., Patla, A. E., and Prince, F. (1998). The initiation of gait in young, elderly, and Parkinson's disease subjects. *Gait and Posture,* 8(1), 8-14.

Harold, L., Klawans, A. H. and Barr, A. (1990). Levodopa-induced dyskinesia. In: G. M. Stern (Ed.), *Parkinson's Disease* (1 ed., pp. 537-540). Baltimore The Johns Hopkins University Press.

Harrington, D. L., Haaland, K. Y. and Hermanowicz, N. (1998). Temporal processing in the basal ganglia. *Neuropsychology,* 12(1), 3-12.

Hase, K. and Stein, R. B. (1998). Analysis of rapid stopping during human walking. *Journal of Neurophysiology,* 80(1), 255-261.

Hashimoto, T., Elder, C. M., Okun, M. S., Patrick, S. K., and Vitek, J. L. (2003). Stimulation of the subthalamic nucleus changes the firing pattern of pallidal neurons. *Journal of Neuroscience,* 23 (5), 1916-1923.

Hass, C., Buckley, T., Pitsikoulis, C., and Barthelemy, E. (2012). Progressive resistance training improves gait initiation in individuals with Parkinson's disease. *Gait and Posture,* 35, 669-673.

Hausdorff, J. M., Cudkowicz, M. E., Firtion, R., Wei, J. Y., and Goldberger, A. L. (1998). Gait variability and basal ganglia disorders: stride-to-stride variations of gait cycle timing in Parkinson's disease and Huntington's disease. *Movement Disorders,* 13(3), 428-437.

Hausdorff, J. M., Mitchell, S. L., Firtion, R., Peng, C. K., Cudkowicz, M. E., Wei, J. Y., and Goldberger, A. L. (1997). Altered fractal dynamics of gait: reduced stride-interval correlations with aging and Huntington's disease. *Journal of Applied Physiology,* 82(1), 262-269.

Hausdorff, J. M., Rios, D. A. and Edelberg, H. K. (2001). Gait variability and fall risk in community-living older adults: a 1-year prospective study. *Archives of Physical Medicine and Rehabilitation,* 82 (8), 1050-1056.

Hausdorff, J. M., Schaafsma, J. D., Balash, Y., Bartels, A. L., Gurevich, T., and Giladi, N. (2003). Impaired regulation of stride variability in Parkinson's disease subjects with freezing of gait. *Experimental Brain Research,* 149(2), 187-194.

Hely, M. A., Morris, J. G., Reid, W. G., and Trafficante, R. (2005). Sydney Multicenter Study of Parkinson's disease: non-L-dopa-responsive problems dominate at 15 years. *Movement Disorders,* 20(2), 190-199.

Hodges, J. R., Salmon, D. P. and Butters, N. (1990). Differential impairment of semantic and episodic memory in Alzheimer's and Huntington's diseases: a controlled propsective study. *Journal of Neurology, Neurosurgery, and Psychiatry,* 53, 1089-1095.

Hoehn, M., Yahr, M. (1967). Parkinsonism: onset, progression, and mortality. *Neurology,* 17 (5), 427-442.

Holford, N. and Nutt, J. G. (2008). Disease progression, drug action and Parkinson's disease: why time cannot be ignored. *European Journal of Clinical Pharmacology,* 64(2), 207-216. doi: 10.1007/ s00228-007-0427-9

Hondzinski, J. (2012). Specificity of training, not the only therapy option for Parkinson's patients. *Journal of Novel Physiotherapies,* 2(6). doi: http://dx.doi.org/10.4172/2165-7025.1000e119

Hoover, J. E. and Strick, P. L. (1993). Multiple output channels in the basal ganglia. *Science,* 259(5096), 819-821.

Huang, X., Mahoney, J., Lewis, M., Du, G., Piazza, S., and Cusumano, J. (2012). Booth coordination and symmetry of arm swing are reduced in Parkinson's disease. *Gait and Posture,* 35, 373-377.

Hughes, A. J., Daniel, S. E., Blankson, S., and Lees, A. J. (1993). A clinicopathologic study of 100 cases of Parkinson's disease. *Archives of Neurology,* 50, 140-148.

Iansek, R. (1999). Interdisciplinary Rehabilitation in Parkinson's Disease. In: G. M. Stern (Ed.), *Parkinson's Disease: Advances in Neurology* (Vol. 80, pp. 557). Philadelphia: Lippincott Williams and Wilkins.

Inglis, W. L., Allen, L. F., Whitelaw, R. B., Latimer, M. P., Brace, H. M., and Winn, P. (1994). An investigation into the role of the pedunculopontine tegmental nucleus in the mediation of locomotion and orofacial stereotypy induced by d-amphetamine and apomorphine in the rat. *Neuroscience,* 58(4), 817-833.

Inglis, W. L. and Winn, P. (1995). The pedunculopontine tegmental nucleus: where the striatum meets the reticular formation. *Progress in Neurobiology,* 47(1), 1-29.

Inzelberg, R., Schechtman, E. and Hocherman, S. (2008). Visuo-motor coordination deficits and motor impairments in Parkinson's disease. *PLoS ONE,* 3(11). doi: 10.1371/journal. pone.0003663

Jain, S., Siegle, G. J., Gu, C., Moore, C. G., Ivanco, L. S., Jennings, J. R., ... Greenamyre, J. T. (2011). Autonomic insufficiency in pupillary and cardiovascular systems in parkinson's disease. *Parkinsonism and Related Disorders,* 17 (2), 119-122. doi: 10.1016/j.parkreldis.2010.11.005

Jankovic, J. (2008). Parkinson's disease: clinical features and diagnosis. *Journal of Neurology, Neurosurgery, and Psychiatry,* 79, 368-376. doi: 10.1136/jnnp.2007.131045

Jenkins, M. E., Almeida, Q. J., Spaulding, S. J., van Oostveen, R. B., Holmes, J. D., Johnson, A. M., and Perry, S. D. (2009). Plantar cutaneous sensory stimulation improves single-limb support time, and EMG activation patterns among individuals with Parkinson's disease. *Parkinsonism and Related Disorders,* 15, 697-702.

Johnsen, E., Sunde, N., Mogensen, P., and Ostergaard, K. (2010). MRI verified STN stimulation site -- gait improvement and clinical outcome. *European Journal of Neurology,* 17, 746-753. doi: 10.1111/j.1468-1331.2010.02962.x

Kadivar, Z., Corcos, D. M., Foto, J., and Hondzinski, J. M. (2011). Effect of step training and rhythmic auditory stimulation on functional performance in Parkinson patients. *Neurorehabilitation and Neural Repair,* 25(7), 626-635. doi: 10.1177/1545968311401627 1545968 311401627 [pii]

Kelland, M. D. and Asdourian, D. (1989). Pedunculopontine tegmental nucleus-induced inhibition of muscle activity in the rat. *Behavioural Brain Research,* 34(3), 213-234.

Kelly, V., Israel, S., Samii, A., Slimp, J., Goodkin, R., and Shumway-Cook, A. (2010). Assessing the effects of subthalamic nucleus stimulation on gait and mobility in people with Parkinson disease. *Disability and Rehabilitation,* 32(11), 929-936.

Koller, W., Pahwa, R., Busenbark, K., Hubble, J., Wilkinson, S., Lang, A., ... Olanow, C. W. (1997). High-frequency unilateral thalamic stimulation in the treatment of essential and parkinsonian tremor. *Annals of Neurology,* 42(3), 292-299.

Konczak, J., Corcos, D., Horak, F., Poizner, H., Shapiro, M., Tuite, P., Volkmann, J., and Maschke, M. (2009). Proprioception and motor control in Parkinson's disease. *Journal of Motor Behavior,* 41(6), 543-552.

Kulisevsky, J. (2000). Role of dopamine in learning and memory: implications for the treatment of cognitive dysfunction in patients with Parkinson's disease. *Drugs and Aging,* 16(5), 365-379.

Kumar, R., Lozano, A. M., Kim, Y. J., Hutchison, W. D., Sime, E., Halket, E., and Lang, A. E. (1998). Double-blind evaluation of subthalamic nucleus deep brain stimulation in advanced Parkinson's disease. *Neurology,* 51(3), 850-855.

Lai, Y. Y. and Siegel, J. M. (1990). Muscle tone suppression and stepping produced by stimulation of midbrain and rostral pontine reticular formation. *Journal of Neuroscience,* 10(8), 2727-2734.

Lamberti, P., Armenise, S., Castaldo, V., de Mari, M., Iliceto, G., Tronci, P., and Serlenga, L. (1997). Freezing gait in Parkinson's disease. *European Neurology,* 38(4), 297-301.

Lee, M., Kim, H. and Lyoo, C. (2005). "Off" gait freezing and temporal discrimination threshold in patients with Parkinson disease. *Neurology,* 64(4), 670-674.

Levangie, P. K. and Norkin, C. C. (2001). *Joint Structure and Function* (3 ed.): Jaypee Borthers Medical Publishers.

Lewis, G. N., Byblow, W. D. and Walt, S. E. (2000). Stride length regulation in Parkinson's disease: the use of extrinsic, visual cues. *Brain,* 123 (Pt 10), 2077-2090.

Lim, I., van Wegen, E., Jones, D., Rochester, L., Nieuwboer, A., Willems, A., ... Kwakkel, G. (2010). Does cueing training improve physical activity in patients with Parkinson's disease? *Neurorehabilitation and Neural Repair,* 24, 469-477.

Limousin, P., Speelman, J. D., Gielen, F., and Janssens, M. (1999). Multicentre European study of thalamic stimulation in parkinsonian and essential tremor. *Journal of Neurology, Neurosurgery, and Psychiatry,* 66(3), 289-296.

Lord, S., Archibald, N., Mosimann, U., Burn, D., and Rochester, L. (2012). Dorsal rather than ventral visual pathways discriminate freezing status in Parkinson's disease. *Parkinsonism and Related Disorders,* 18, 1094-1096.

Luessi, F., Mueller, L., Breimhorst, M., and Vogt, T. (2012). Influence of visual cues on gait in Parkinson's disease during treadmill walking at multiple velocities. *Journal of the Neurological Sciences,* 314, 78-82. doi: 10.1016/j.jns.2011.10.027

Lundin-Olsson, L., Nyberg, L. and Gustafson, Y. (1997). "Stops walking when talking" as a predictor of falls in elderly people. *The Lancet,* 349, 617.

MacKay-Lyons, M. (1998). Variability in spatiotemporal gait characteristics over the course of the L-dopa cycle in people with advanced Parkinson disease. *Physical Therapy,* 78(10), 1083-1094.

Maidan, I., Plotnik, M., Mirelman, A., Weiss, A., Giladi, N., and Hausdorff, J. (2010). Heart rate changes during freezing of gait in patients with Parkinson's disease. *Movement Disorders,* 25(14), 2346-2354.

Martin, M., Shinberg, M., Kuchibhatla, M., Ray, L., Carollo, J. J., and Schenkman, M. L. (2002). Gait initiation in community-dwelling adults with Parkinson disease: comparison with older and younger adults without the disease. *Physical Therapy,* 82(6), 566-577.

Matinolli, M., Korpelainen, J., Korpelainen, R., Sotaniemi, K., and Myllyla, V. (2009). Orthostatic hypotension, balance, and falls in Parkinson's disease. *Movement Disorders,* 24(5), 745-751.

Mbourou, G. A., Lajoie, Y. and Teasdale, N. (2003). Step length variability at gait initiation in elderly fallers and non-fallers, and young adults. *Gerontology,* 49(1), 21-26. doi: ger 49021 [pii]

McDonald, W., Richard, I., DeLong, M. (2003). Prevalence, etiology, and treatment of depression in Parkinson's disease. *Biological Psychiatry,* 54(3), 363-375.

McIntosh, G. C., Brown, S. H., Rice, R. R., and Thaut, M. H. (1997). Rhythmic auditory-motor facilitation of gait patterns in patients with Parkinson's disease. *Journal of Neurology, Neurosurgery, and Psychiatry,* 62(1), 22-26.

McIntyre, C. C., Savasta, M., Kerkerian-Le Goff, L., and Vitek, J. L. (2004). Uncovering the mechanism(s) of action of deep brain stimulation: activation, inhibition, or both. *Clinical Neurophysiology,* 115(6), 1239-1248.

Meara, J. and Koller, W. C. (2000). *Parkinson's disease and parkinsonism in the elderly* J. Meara and W. C. Koller (Eds.), doi:051100608

Mena-Segovia, J., Bolam, J. P. and Magill, P. J. (2004). Pedunculopontine nucleus and basal ganglia: distant relatives or part of the same family? *Trends in Neurosciences,* 27(10), 585-588.

Menz, H. B., Latt, M. D., Tiedemann, A., Mun San Kwan, M., and Lord, S. R. (2004). Reliability of the GAITRite walkway system for the quantification of temporo-spatial parameters of gait in young and older people. *Gait and Posture,* 20(1), 20-25.

Merritt, H. H. (1963). Paralysis Agitans. In: H. H. Merritt (Ed.), *A Textbook of Neurology* (3 ed., pp. 475). Philadelphia: Lea and Febiger.

Middleton, F. A. and Strick, P. L. (1997). New concepts about the organization of basal ganglia output. *Advances in Neurology,* 74, 57-68.

Mileykovskiy, B. Y., Kiyashchenko, L. I., Kodama, T., Lai, Y. Y., and Siegel, J. M. (2000). Activation of pontine and medullary motor inhibitory regions reduces discharge in neurons located in the locus coeruleus and the anatomical equivalent of the midbrain locomotor region. *Journal of Neuroscience,* 20(22), 8551-8558.

Mille, M., Hilliard, J., Martinez, K., Simuni, T., Zhang Y., and Rogers, M. (2009). Short-term effects of posture-assisted step training on rapid step initiation in Parkinson's disease. *Journal of Neurologic Physical Therapy,* 33, 88-95.

Miller, R. A., Thaut, M. H., McIntosh, G. C., and Rice, R. R. (1996). Components of EMG symmetry and variability in parkinsonian and healthy elderly gait. *Electroencephalography and Clinical Neurophysiology,* 101(1), 1-7.

Mink, J. W. (2007). Functional Organization of the Basal Ganglia. In: J. Jankovic and T. Eduardo (Eds.), *Parkinson's Disease and Movement Disorder* (5 ed., pp. 1-3). Philadelphia: Lippincott Williams and Wilkins.

Mirelman, A., Maidan, I., Deutsch, J., Giladi, N., and Hausdorff, J. (2011). Virtual reality for gait training: can it induce motor learning to enhance complex walking and reduce fall risk in patients with Parkinson's disease? *Journals of Gerontology: Biological Sciences and Medical Sciences,* 66, 234-240.

Mitoma, H., Hayashi, R., Yanagisawa, N., and Tsukagoshi, H. (2000). Characteristics of parkinsonian and ataxic gaits: a study using surface electromyograms, angular displacements and floor reaction forces. *Journal of the Neurological Sciences,* 174(1), 22-39.

Moes, E. and Lombardi, K. (2009). The relationship between contrast sensitivity, gait, and reading speed in Parkinson's disease. *Neuropsychology, development, and cognition: Aging, neuropsychology and cognition,* 16(2), 121-132. doi: 10.1080/1382558080 2233418

Monakow, K. H., Akert, K. and Kunzle, H. (1978). Projections of the precentral motor cortex and other cortical areas of the frontal lobe to the subthalamic nucleus in the monkey. *Experimental Brain Research,* 33(3-4), 395-403.

Morris, M. E. (2000). Movement disorders in people with Parkinson disease: a model for physical therapy. *Physical Therapy,* 80(6), 578-597.

Morris, M. E., Iansek, R., Matyas, T. A., and Summers, J. J. (1994a). Ability to modulate walking cadence remains intact in Parkinson's disease. *Journal of Neurology, Neurosurgery, and Psychiatry,* 57 (12), 1532-1534.

Morris, M. E., Iansek, R., Matyas, T. A., and Summers, J. J. (1994b). The pathogenesis of gait hypokinesia in Parkinson's disease. *Brain,* 117 (Pt 5), 1169-1181.

Morrish, P. K., Sawle, G. V. and Brooks, D. J. (1996). An [18F]dopa-PET and clinical study of the rate of progression in Parkinson's disease. *Brain,* 119 (Pt 2), 585-591.

Murray, M. P., Sepic, S. B., Gardner, G. M., and Downs, W. J. (1978). Walking patterns of men with parkinsonism. *American Journal of Physical Medicine and Rehabilitation,* 57(6), 278-294.

Nagasaki, H., Itoh, H., Hashizume, K., Furuna, T., Maruyama, H., and Kinugasa, T. (1996). Walking patterns and finger rhythm of older adults. *Perceptual and Motor Skills,* 82(2), 435-447.

Nakamura, T., Bronstein, A., Lueck, C., Marsden, C., Rudge, P. (1994). Vestibular, cervical and visual remembered saccades in Parkinson's disease. *Brain,* 117(6), 1423-1432.

Nambu, A. (2004). A new dynamic model of the cortico-basal ganglia loop. *Progress in Brain Research,* 143, 461-466.

Nambu, A. (2005). A new approach to understand the pathophysiology of Parkinson's disease. *Journal of Neurology,* 252 Suppl. 4, IV1-IV4.

Nambu, A., Tokuno, H., Hamada, I., Kita, H., Imanishi, M., Akazawa, T., ... Hasegawa, N. (2000). Excitatory cortical inputs to pallidal neurons via the subthalamic nucleus in the monkey. *Journal of Neurophysiology,* 84(1), 289-300.

Narabayashi, H. (1993). Three types of akinesia in the progressive course of Parkinson's disease. *Advances in Neurology,* 60, 18-24.

Naugle, K., Hass, C., Bowers, D., and Janelle, C. (2012). Emotional state affects gait initiation in individuals with Parkinson's disease. *Cognitive, Affective, and Behavioral Neuroscience,* 12, 207-219.

Nauta, H. J. (1979). A proposed conceptual reorganization of the basal ganglia and telencephalon. *Neuroscience,* 4(12), 1875-1881.

Nieuwboer, A., Baker, K., Willems, A. M., Jones, D., Spildooren, J., Lim, I., ... Rochester, L. (2009). The Short-Term Effects of Different Cueing Modalities on Turn Speed in People With Parkinson's Disease. *Neurorehabilitation and Neural Repair,* 23, 831-836. doi: 154596830933 7136 [pii] 10.1177/1545968309337136

Nieuwboer, A., Dom, R., De Weerdt, W., Desloovere, K., Fieuws, S., and Broens-Kaucsik, E. (2001). Abnormalities of the spatiotemporal characteristics of gait at the onset of freezing in Parkinson's disease. *Movement Disorders,* 16(6), 1066-1075.

Nieuwboer, A., Dom, R., De Weerdt, W., Desloovere, K., Janssens, L., and Stijn, V. (2004). Electromyographic profiles of gait prior to onset of freezing episodes in patients with Parkinson's disease. *Brain,* 127(Pt 7), 1650-1660.

Niu, L., Ji, L-Y., Li, J-M., Zhao, D-S., Guang, G., Liu, W-P., ... Ji, X-T. (2012). Effect of bilateral deep brain stimulation of the subthalamic nucleus on freezing of gait in Parkinson's disease. *The Journal of International Medical Research*, 40, 1108-1113.

Nolano, M., Provitera, V., Estraneo, A., Selim, M., Caporaso, G., Stancanelli, A., ... Santoro, L. (2008). Sensory deficit in Parkinson's disease: evidence of a cutaneous denervation. *Brain,* 131, 1903-1911. doi: 10.1093/brain/ awn102

Novak, P. and Novak, V. (2006). Effect of step-synchronized vibration stimulation of soles on gait in Parkinson's disease: a pilot study. *Journal of NeuroEngineering and Rehabilitation,* 3, 9.

Nussbaum, R. L. and Ellis, C. E. (2003). Alzheimer's disease and Parkinson's disease. *The New England Journal of Medicine,* 348, 1356-1364.

O'Boyle, D. J., Freeman, J. S. and Cody, F. W. (1996). The accuracy and precision of timing of self-paced, repetitive movements in subjects with Parkinson's disease. *Brain,* 119 (Pt 1), 51-70.

O'Suilleabhain, P., Bullard, J., Dewey, R. (2001). Proprioception in Parkinson's disease is acutely depressed by dopaminergic medications. *Journal of Neurology, Neurosurgery, and Psychiatry,* 71, 607-610.

Obeso, J. A., Rodriguez-Oroz, M. C., Rodriguez, M., Lanciego, J. L., Artieda, J., Gonzalo, N., and Olanow, C. W. (2000). Pathophysiology of the basal ganglia in Parkinson's disease. *Trends in Neurosciences,* 23(10 Suppl.), S8-19.

Pahwa, R., Lyons, K. E., Wilkinson, S. B., Troster, A. I., Overman, J., Kieltyka, J., and Koller, W. C. (2001). Comparison of thalamotomy to deep brain stimulation of the thalamus in essential tremor. *Movement Disorders,* 16(1), 140-143.

Pahwa, R., Lyons, K., Koller, W., and Jankovic, J. (2007). Pathophysiology and assessment of Parkinsonian symptoms and signs. In: R. Pahwa, Lyons, K., Koller, W. (Ed.), *Handbook of Parkinson's disease* (pp. 79-104). New York: Taylor and Francis Group, LLC.

Pai, Y. C. and Patton, J. (1997). Center of mass velocity-position predictions for balance control. *Journal of Biomechanics,* 30(4), 347-354.

Pandya, M., Kubu, C. S., Giroux, M. L. (2008). Parkinson disease: not just a movement disorder. *Cleveland Clinic Journal of Medicine,* 75(12), 856-864. doi: 10.3949/ccjm. 75a.07005

Parati, E. A., Fetoni, V., Geminiani, G. C., Soliveri, P., Giovannini, P., Testa, D., ... Girotti, F. (1993). Response to L-DOPA in multiple system atrophy. *Clinical Neuropharmacology,* 16, 139-144.

Parent, A. and Hazrati, L. N. (1995). Functional anatomy of the basal ganglia. I. The cortico-basal ganglia-thalamo-cortical loop. *Brain Research Reviews,* 20(1), 91-127.

Pastor, M. A., Artieda, J., Jahanshahi, M., and Obeso, J. A. (1992). Time estimation and reproduction is abnormal in Parkinson's disease. *Brain,* 115 Pt 1, 211-225.

Pedersen, S. W., Oberg, B., Larsson, L. E., and Lindval, B. (1997). Gait analysis, isokinetic muscle strength measurement in patients with Parkinson's disease. *Scandinavian Journal of Rehabilitation Medicine,* 29(2), 67-74.

Piper, M., Abrams, G. and Marks Jr., W. (2005). Deep brain stimulation for the treatment of Parkinson's disease: overview and impact on gait and mobility. *NeuroRehabilitation,* 20, 223-232.

Potter-Nerger, M., Reich, M. M., Colebatch, J. G., Deuschl, G., and Volkmann, J. (2012). Differential effect of dopa and subthalamic stimulation on vestibular activity in Parkinson's disease. *Movement Disorders,* 27(10), 1268-1275. doi: 10.1002/mds.25061

Pratorius, B., Kimmeskamp, S. and Milani, T. (2003). The sensitivity of the sole of the foot in patients with Morbus Parkinson. *Neuroscience Letters,* 346, 173-176.

Prediger, R., Matheus, F., Schwarzbold, M., Lima, M., Vital, M. (2012). Anxiety in Parkinson's disease: a critical review of experimental and clinical studies. *Neuropharmacology,* 62(1), 115-124. doi: 10.1016/j.neuropharm.2011.08.039

Pursiainen, V., Haapaniemi, T., Korpelainen, J., Sotaniemi, K., and Myllyla, V. (2007). Sweating in Parkinsonian patients with wearing-off. *Movement Disorders,* 22, 828-832.

Ransmayr, G. (2011). Physical, occupational, speech and swallowing therapies and physical exercise in Parkinson's disease. *Journal of Neural Transmission,* 118, 773-781. doi: 10.1007/s00702-011-0622-9

Reichert, W., Doolittle, J. and McDowell, F. (1982). Vestibular dysfunction in Parkinson disease. *Neurology,* 32(10), 1133-1138.

Ridgel, A. L., Vitek, J. L. and Alberts, J. L. (2009). Forced, Not Voluntary, Exercise Improves Motor Function in Parkinson's Disease Patients. *Neurorehabilitation and Neural Repair.* doi: 1545968308328726 [pii] 10.1177/1545968308328726

Robertson, C. and Flowers, K. A. (1990). Motor set in Parkinson's disease. *Journal of Neurology, Neurosurgery, and Psychiatry,* 53(7), 583-592.

Robichaud, J. A., Pfann, K. D., Vaillancourt, D. E., Comella, C. L., and Corcos, D. M. (2005). Force control and disease severity in parkinson's disease. *Movement Disorders,* 20 (4), 441-450. doi: 10.1002/mds.20350

Rocchi, L., Carlson-Kuhta, P., Chiari, L., Burchiel, K., Hogarth, P., and Horak, F. (2012). Effects of deep brain stimulation in the subthalamic nucleus or globus pallidus internus on step initiation in Parkinson disease. *Journal of Neurosurgery,* 117(6), 1141-1149. doi: 10.3171/ 2012.8.JNS112006

Rochester, L., Baker, K., Hetherington, V., Jones, D., Willems, A., Kwakkel, G., ... Nieuwboer, A. (2010). Evidence for motor learning in Parkinson's disease: acquisition, automaticity and retention of cued gait performance after training with external rhythmical cues. *Brain Research,* 1319. doi: 10.1016/j.brainres.2010.01.001

Rochester, L., Burn, D., Woods, G., Godwin, J., and Nieuwboer, A. (2009). Does auditory rhythmical cueing improve gait in people with Parkinson's disease and cognitive impairment? A feasibility study. *Movement Disorders,* 24(6).

Rodriguez, R. L., Fernandez, H. H., Haq, I., and Okun, M. S. (2007). Pearls in patient selection for deep brain stimulation. *Neurologist,* 13(5), 253-260.

Rosin, R., Topka, H., Dichgans, J. (1997). Gait initiation in Parkinson's disease. *Movement Disorders,* 12(5), 682-690.

Rossi-Izquierdo, M., Soto-Varela, A., Santos-Perez, S., Sesar-Ignacio, A., and Labella-Caballero, T. (2009). Vestibular rehabilitation with computerised dynamic posturography in patients with Parkinson's disease: improving balance impairment. *Disability and Rehabilitation,* 31(23), 1907-1916.

Russchen, F. T., Bakst, I., Amaral, D. G., and Price, J. L. (1985). The amygdalostriatal projections in the monkey. An anterograde tracing study. *Brain Research,* 329(1-2), 241-257.

Rye, D. B., Lee, H. J., Saper, C. B., and Wainer, B. H. (1988). Medullary and spinal efferents of the pedunculopontine tegmental nucleus and adjacent mesopontine tegmentum in the rat. *Journal of Comparative Neurology,* 269(3), 315-341.

Sakakibara, R., Kishi, M., Ogawa, E., Tateno, F., Uchiyama, T., Yamamoto, T., and Yamanishi, T. (2011). Bladder, bowel, and sexual dysfunction in Parkinson's disease. *Parkinson's Disease*. doi: 10.4061/2011/924605

Schaafsma, J. D., Giladi, N., Balash, Y., Bartels, A. L., Gurevich, T., and Hausdorff, J. M. (2003). Gait dynamics in Parkinson's disease: relationship to Parkinsonian features, falls and response to levodopa. *Journal of the Neurological Sciences,* 212(1-2), 47-53.

Schrag, A., Sampaio, C., Counsell, N., Poewe, W. (2006). Minimal clinically important change on the unified Parkinson's disease rating scale. *Movement Disorders,* 21(8), 1200-1207.

Schubert, M., Prokop, T., Brocke, F., and Berger, W. (2005). Visual kinesthesia and locomotion in Parkinson's disease. *Movement Disorders,* 20(2), 141-150.

Sellby, G. (1990). Clinical features. In: G. M. Stern (Ed.), *Parkinson's Disease* (1 ed., pp. 337-353). Baltimore: The Johns Hopkins University Press.

Sheridan, M. R. and Flowers, K. A. (1990). Movement variability and bradykinesia in Parkinson's disease. *Brain,* 113 (Pt 4), 1149-1161.

Shibasaki, H., Tsuji, S., Kuroiwa, Y. (1979). Oculomotor abnormalities in Parkinson's disease. *Archives of Neurology,* 36(6).

Shik, M. L., Severin, F. V. and Orlovskii, G. N. (1966). [Control of walking and running by means of electric stimulation of the midbrain]. *Biofizika,* 11(4), 659-666.

Skinner, R. D. and Garcia-Rill, E. (1984). The mesencephalic locomotor region (MLR) in the rat. *Brain Research,* 323(2), 385-389.

Skinner, R. D., Kinjo, N., Ishikawa, Y., Biedermann, J. A., and Garcia-Rill, E. (1990). Locomotor projections from the pedunculopontine nucleus to the medioventral medulla. *Neuroreport,* 1(3-4), 207-210.

Smulders, K., Esselink, R.A.J., Weiss, A., Kessels, R.P.C., Geurts, A.C.H., and Bloem, B.R. (2012). *Assessment of dual tasking has no clinical value for fall prediction in Parkinson's disease*. 259(2), 1840-1847. doi: DOI 10.1007/s00415-012-6419-4

Sparrow, W. A. and Tirosh, O. (2005). Gait termination: a review of experimental methods and the effects of ageing and gait pathologies. *Gait and Posture,* 22(4), 362-371.

Stack, E. L., Ashburn, A. M. and Jupp, K. E. (2006). Strategies used by people with Parkinson's disease who report difficulty turning. *Parkinsonism and Related Disorders,* 12(2), 87-92.

Starr, A., Angel, R., Yeates, H. (1969). Visual suppression during smooth following and saccadic eye movements. *Vision Research,* 9(1), 195-197.

Stelmach, G. E., Teasdale, N., Phillips, J., and Worringham, C. J. (1989). Force production characteristics in Parkinson's disease. *Experimental Brain Research,* 76(1), 165-172.

Takakusaki, K., Habaguchi, T., Ohtinata-Sugimoto, J., Saitoh, K., and Sakamoto, T. (2003). Basal ganglia efferents to the brainstem centers controlling postural muscle tone and locomotion: a new concept for understanding motor disorders in basal ganglia dysfunction. *Neuroscience,* 119(1), 293-308.

Tan, T., Almeida, Q. and Rahimi, F. (2011). Proprioceptive deficits in Parkinson's disease patients with freezing of gait. *Neuroscience,* 19, 746-752.

Tedroff, J., Ekesbo, A., Rydin, E., Langstrom, B., Hagberg, G. (1999). Regulation of dopaminergic activity in early Parkinson's disease. *Annals of Neurology,* 26(3), 359-365.

Thaut, M., McIntosh, G., Rice, R., Miller, R., Rathbun, J., and Brault, J. (1996). Rhythmic auditory stimulation in gait training for Parkinson's disease patients. *Movement Disorders,* 11, 193-200.

Thobois, S., Delamarre-Damier, F. and Derkinderen, P. (2005). Treatment of motor dysfunction in Parkinson's disease: an overview. *Clinical Neurology and Neurosurgery,* 107(4), 269-281.

Tolosa, E. and Kalzenschlager, R. (2007). Pharmalogical management of Parkinson's Disease. In: J. Jankovic and E. Tolosa (Eds.), *Parkinson's Disease and Movement Disorder* (5 ed., pp. 110-145). Philadelphia: Lippincott Williams and Wilkins.

Toxopeus, C. M., de Jong, B. M., Valsan, G., Conway, B. A., van der Hoeven, J. H., Leenders, K. L., Maurits, N. M. (2011). Impairment of gradual muscle adjustment during wrist circumduction in Parkinson's disease. *PLoS One,* 6(9), e24572. doi: 10.1371/ journal.pone.0024572

Tseng, I., Jeng, C. and Yuan, R. (2012). Comparisons of forward and backward gait between poorer and better attention capabilities in early Parkinson's disease. *Gait and Posture,* 36, 367-371.

Uc, E., Rizzo, M., Anderson, S., Oian, S., Rodnitzky, R., and Dawson, J. (2005). Visual dysfunction in Parkinson disease without dementia. *Neurology,* 65(12), 1907-1913.

Ueno, E., Yanagisawa, N. and Takami, M. (1993). Gait disorders in parkinsonism. A study with floor reaction forces and EMG. *Advances in Neurology,* 60, 414-418.

Van Den Eeden, S. K., Tanner, C. M., Bernstein, A. L., Fross, R. D., Leimpeter, A., Bloch, D. A., and Nelson, L. M. (2003). Incidence of Parkinson's disease: variation by age, gender, and race/ethnicity. *American Journal of Epidemiology,* 157(11), 1015-1022.

Vercruysse, S., Devos, H., Munks, L., Spildooren, J., Vandenbossche, J., Vandenberghe, W., Nieuwboer, A., and Heremans, E. (2012). Explaining freezing of gait in Parkinson's disease: motor and cognitive determinants. *Movement Disorders,* 27(13), 1644-1651.

Vick, N. A. (1976). Disorders of the Basal Ganglia and Thalamus. In: A. Gabell (Ed.), *Grinker's Neurology* (7 ed., pp. 333-334). Springfield: Charles C Thomas.

Vieregge, P., Stolze, H., Klein, C., and Heberlein, I. (1997). Gait quantitation in Parkinson's disease--locomotor disability and correlation to clinical rating scales. *Journal of Neural Transmission,* 104(2-3), 237-248.

Vitorio, R., Lirani-Silva, E., Barbieri, F., Raile, V., Batistela, R., Stella, F., Bucken Gobbi, L. (2012). The role of vision in Parkinson's disease locomotion control: free walking task. *Gait and Posture,* 35, 175-179.

Volkmann, J., Allert, N., Voges, J., Sturm, V., Schnitzler, A., and Freund, H. J. (2004). Long-term results of bilateral pallidal stimulation in Parkinson's disease. *Annals of Neurology,* 55(6), 871-875.

Watson, G. S., Cholerton, B. A., Gross, R. G., Weintraub, D., Zabetian, C. P., Trojanowski, J. Q., . . . Leverenz, J. B. (2013). Neuropsychologic assessment in collaborative Parkinson's disease research: A proposal from the National Institute of Neurological Disorders and Stroke Morris K. Udall Centers of Excellence for Parkinson's Disease Research at the University of Pennsylvania and the University of Washington. *Alzheimer's and Dementia, 9*(5), 609-614. doi: 10.1016/j.jalz.2012.07.006Webster, K. (Ed.). (1990). *The functional anatomy of the basal ganglia* (1 ed.): The John Hopkins University Press.

Weinrich, M. and Bhatia, R. (1986). Abnormal eye-head coordination in Parkinson's disease patients after administration of levodopa: a possible substrate of levodopa-induced dyskinesia. *Journal of Neurology, Neurosurgery, and Psychiatry,* 49, 785-790.

Welter, M. L., Houeto, J. L., Bonnet, A. M., Bejjani, P. B., Mesnage, V., Dormont, D., ... Pidoux, B. (2004). Effects of high-frequency stimulation on subthalamic neuronal activity in parkinsonian patients. *Archives of Neurology,* 61(1), 89-96.

White, O., Saint-Cyr, J., Tomlinson, R., Sharpe, J. (1983). Ocular motor deficits in Parkinson's disease. II. Control of the saccadic and smooth pursuit systems. *Brain,* 106 (3), 571-587.

Willems, A., Nieuwboer, A., Chavret, F., Desloovere, K., Dom, R., Rochester, L., … Jones, D. (2007). Turning in Parkinson's disease patients and controls: the effect of auditory cues. *Movement Disorders,* 22(13), 1871-1878.

Williams-Gray, C., Foltynie, T., Brayne, C., Robbins, T., and Barker R. (2007). Evolution of cognitive dysfunction in an incident Parkinson's disease cohort. *Brain,* 130, 1787-1798. doi: 10.1093/brain/awm111

Wing, A. M., Keele, S. and Margolin, D. I. (1984). Motor disorder and the timing of repetitive movements. *Annals of the New York Academy of Sciences,* 423, 183-192.

Winogrodszka, A. W., R. C., Booij, J., Wolters, E. C. (2005). Rigidity and bradykinesia reduce interlimb coordination in Parkinsonian gait. *Archives of Physical Medicine and Rehabilitation,* 86(2), 183-189.

Wright, W., Gurfinkel, V., King, L., Nutt, J., Cordo, P., Horak, F. (2010). Axial kinesthesia is impaired in Parkinson's disease: effects of levodopa. *Experimental Neurology,* 225, 202-209.

Yekhlef, F., Ballan, G., Macia, F., Delmer, O., Sourgen, C., and Tison, F. (2003). Routine MRI for the differential diagnosis of Parkinson's disease, MSA, PSP, and CBD. *Journal of Neural Transmission,* 110(2), 151-169.

Yogev-Seligman, G., Giladi, N., Brozgol, M., and Hausdorff, J. (2012). A training program to improve gait while dual tasking in patients with Parkinson's disease: apilot study. *Archives of Physical Medicine and Rehabilitation,* 93, 176-181.

Yogev, G., Plotnik, M., Peretz, C., Giladi, N., and Hausdorff, J. (2007). Gait assymetry in patients with Parkinson's disease and elderly fallers: when does the bilateral coordination of gait require attention? *Experimental Brain Research,* 177, 336-346.

In: Gait Biometrics
Editors: Li Li and Matthew Holmes

ISBN: 978-1-62948-381-8

Chapter VI

Gait Analysis of Two Incomplete Spinal Cord Syndromes: Central Cord and Brown-Sequard*

S. Pérez-Nombela[1], A. Gil-Agudo[1], L. Garcés-Pérez[2] and D. Torricelli[3]

[1] Biomechanics and Technical Aids Unit and Department of Physical Medicine and Rehabilitation, National Spinal Cord Injury Hospital, SESCAM, Toledo, Spain

[2] Biomechanic Institute of Valencia. Valencia, Spain

[3] Bioengineering Group, CSIC, Madrid, Spain

Abstract

The treatment of gait disorders in incomplete spinal cord syndromes has been improved with specific therapies such as functional electrical stimulation (FES), oral or intrathecal baclofen, botulinum toxin and orthoses which are able to return gait capacity or to correct abnormal gait patterns. The combination of the biomechanical gait analysis with these therapeutic possibilities allows better functional outcomes for patients.

Central cord syndrome (CCS) and cord hemi-section, or Brown-Séquard syndrome (BSS), are the two most remarkable syndromes of incomplete SCI. Despite the high percentage of CCS and BSS patients who can walk previous to, or after rehabilitation, there are few studies on the gait characteristics in both syndromes. Describing abnormal gait patterns to be expected due to different neuro-motor disorders, allows for the development of new approaches to treatment and the definition of specific rehabilitation programs.

Patients with incomplete SCI do not just walk at a slower speed due to the need to focus on balance and stability, but also display a series of kinematic alterations such as: a smaller range of movement in the sagittal plane of the knee with limited flexion during

*This work was supported by CONSOLIDER-INGENIO 2010 program HYPER CSD2009-00067. MICINN (Spain).

the oscillation phase; greater abduction of the hip at the initial contact and during the oscillation phase; and a diminished range of joint movements in the ankle.

Ankle kinematics seems to be different between subjects. Patients with CCS show more dorsiflexion. The differences in ankle kinematics had consequences at the knee for each of these patients. To achieve a comparable level of walking and compensate for the deficit, subjects with CCS may have a higher motor score and better control of lower extremities than subjects with BSS.

Nevertheless, not just ROM limitations, but also motor control disorders due to lack of specificity in muscle control, can be responsible for these findings. From a clinical point of view, the reduction of knee flexion at initial contact in patients with CCS is suggested. This strategy could involve lengthening the knee flexor muscles and working on muscle coordination, rather than simply strengthening the muscles. Biomechanical gait analysis can reveal information pertinent to the selection of a specific rehabilitation program for each patient to enhance gait training as well as its therapeutic effects that clinical evaluations alone cannot provide. These different approaches show the importance of gait analysis as a tool to make therapeutic decisions, such as the prescription of orthoses or a surgical procedure, and to evaluate the patient during treatment or after surgical intervention.

1. Introduction

It is estimated that there are between 219,000 and 279,000 persons in the United States now living with the consequences of spinal cord injury (SCI). There is an estimated annual incidence of 30-40 per million population: therefore, approximately 8000-11,000 persons with SCI are being added each year [1]. The percentage of persons with incomplete SCI has increased from 44% in 1973 to 54% in 1990 [2]. With the use of proper bracing and walking aids, many of these individuals with incomplete SCI become functional walkers [3]. Restoring the capacity to walk in patients with incomplete spinal cord injury (SCI) is one of the main goals in rehabilitation programs. But only a few percentage of patients with SCI become again this capability, since most percentage of patients with Spinal Cord Injuries has to use a wheelchair to do their displacements. The lower the level of the injury, the greater possibilities of walking with the help of less assistive devices. The spasticity and injury level determine the pattern of abnormality in gait after spinal cord injury [4]

Nevertheless, incomplete spinal cord injuries are the group in which major gait re-education attempts are focused, specially at ASIA D injuries.

As many as 17 to 41 per cent of spinal cord injuries are reported to be motor incomplete, such that significant motor recovery occurs (Bosch et al., 1971; Bedbrook, 1979; Young and Northrup, 1979). Commonly these patients are totally paralyzed initially but gradually regain voluntary motor strength and functional ambulation.

Walking prognosis usually depends upon muscle power loss, degree of spasticity present, type of lower limb joint deformity developing and availability of treatment. [5,6]

Five distinct syndromes can be categorized within the wide spectrum of Incomplete spinal cord injury: the Central Cord, the Brown-Sequard, the anterior cord, the conus medullaris and the cauda equina syndromes.

Central cord syndrome (CCS) and cord hemi-section or Brown-Séquard syndrome (BSS) are the two most remarkable of them with gait disturbances [1].

Walking ability is important for quality of life and participation in social and economic life, but can be adversely affected by neurological disorders such as spinal cord injury. So the gait rehabilitation and evaluation is one of the most important goals in the treatment of these patients.

Gait of children and adolescents with SCI after interventions [7-9] or the description of the effects of tone-reducing orthotics on walking of an individual after incomplete spinal cord injury [7] have been reported.

However, the specific walking disorders occurring after incomplete SCI have been scarcely described in the literature but a recent study described the disturbances in the gait patterns of children and adolescents with SCI underscoring the importance of gait analysis as a tool to take therapeutic decisions, such as the prescription of orthoses or a surgical procedure, and to evaluate the patient during treatment or after surgical intervention [10]

As complementation to physical examination and direct visual gait analysis, instrumented gait analysis of these patients, including three-dimensional joint range motion (kinematics), estimation of joint forces occurring at the hip, knee and ankle/foot (kinetics), dynamic EMG and energy cost tests will allow us to better understand a gait deficit. Knowledge of the effects of any dynamic contracture on the available redundancy within ipsilateral limb joints, and information about muscle actions obtained by calculating joint forces (including moments and powers) improves overall knowledge of gait cycle abnormality.

However, there are few studies that have focussed on the biomechanics of gait in patients with Incomplete SCI. To date, comparative biomechanical data has only been obtained in such patients for gait aided by one or two walking sticks [11].

However, the need to use biomechanical analyses to evaluate this patient group has been already emphasized [10,12]. Despite the high percentage of CCS and BSS patients who can walk, there are only limited studies on the gait characteristics in both syndromes [12,13]. A recent study compared the gait characteristics of both syndromes by measuring the speed or using specific scales like Walking Index for Spinal Cord Injury (WISCI-II) [14]. But no studies have been found considering or comparing the biomechanical characteristics of both syndromes.

So, the objective of this chapter is to track the information of the gait analysis in SCI patients and should then assist the clinical decision-making process. All this information should then assist the clinical decision-making process.The most important gait disorders of these SCI syndromes will be described in order to provide a better approach to their treatments.

2. Biomechanical Gait Analysis

Several articles describe gait analyses carried out in neuro-motor disorders. These studies provide the basis to describe the type of gait disturbances that can be expected in these groups of patients and serve to define a rehabilitation therapy with realistic goals.

The typical gait analysis laboratory is endowed with kinetic, kinematic and electromyography sensors [15] (Figure 1). Several cameras or receptors are placed around a walkway or a treadmill, which are linked to a computer. The patient has markers located at various points of reference of the body (e.g. iliac spines of the pelvis, ankle malleolus, and the condyles of the knee), or groups of markers applied to half of the body segments. To perform a gait analysis, the patient has to walk down the walkway or the treadmill and the computer calculates the trajectory of each marker in three dimensions. A biomechanic model is applied to calculate the movement of the body segments. This gives a complete breakdown of the movement of each joint.

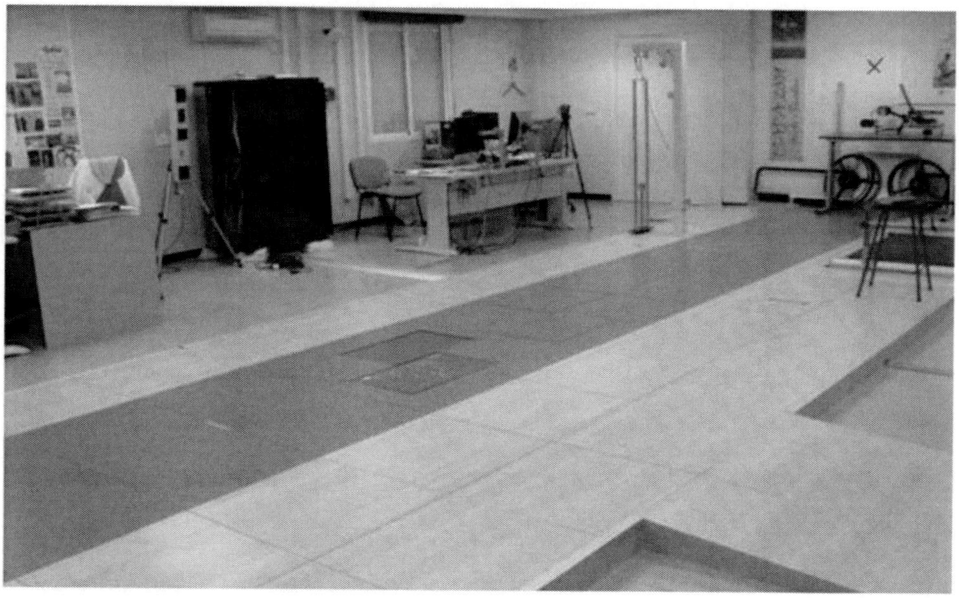

Figure 1. Human movement analysis laboratory.

Kinetics are calculated with the use of force platforms, which measure the ground reaction forces and moments, including the magnitude, direction and location (called the centre of pressure). The spatial distribution of pressures can be measured with podobarography equipment (insoles or platforms).

Inverse dynamics are the algorithms to calculate the net forces and the net moments of force about each joint at every stage of the gait cycle. Using the known dynamics of each body segment, enables the solution of equations based on the Newton–Euler equations of motion.

Electromyography allows to detect the activity and contribution of individual muscles to movement through the electrical activity of muscles. Many gait labs use surface electrodes attached to the skin to detect the electrical activity or electromyography (EMG). The times and magnitude of activation of muscles and their contribution to gait can be measured.

The comparison of the individual analysis with normal kinematic, kinetic, or EMG patterns are used to study specific pathologies, predict the outcome of treatments, or determine the effectiveness of training programs.

Cine film or video recordings using footage from single or multiple cameras can be used to measure joint angles and velocities. This method has been aided by the development of analysis software that greatly simplifies the analysis process and allows for analysis in three dimensions rather than two dimensions only.

Passive marker systems, using reflective markers (typically reflective balls), allows for very accurate measurement of movements using multiple cameras (typically five to twelve cameras), simultaneously. The cameras utilize high-powered strobes (typically red, near infrared or infrared) with matching filters to record the reflection from the markers placed on the body. Markers are located at palpable anatomical landmarks. Based on the angle and time delay between the original and reflected signal, triangulation of the marker in space is possible. Software is used to create three dimensional trajectories from these markers which are subsequently given identification labels. A computer model is then used to compute joint angles from the relative marker positions of the labelled trajectories. These are also used for motion capture in the motion picture industry.

Active marker systems are similar to the passive marker system but use "active" markers (Figure 2). These markers are triggered by the incoming infra red signal and respond by sending out a corresponding signal of their own. This signal is then used to triangulate the location of the marker. The advantage of this system over the passive one is that individual markers work at predefined frequencies and therefore, have their own "identity". This means that no post-processing of marker locations is required, however, the systems tend to be less forgiving for out-of-view markers than the passive systems.

Since there are many parameters that can be obtained from gait analysis, it is necessary to take into account the reliability of measurements in different joint planes. In marker based gait analysis, some of these parameters can be obtained with greater precision (hip and knee ROM in the sagittal plane) than others (such as hip or knee rotation), since a larger movement is measured.

Inertial measurement units (IMUs) provide yet another alternative to obtain kinematic information and these sensors are designed to measure the orientation of an object within a given space. As they provide accurate readings without inherent latency, these sensors are useful for human motion tracking applications. There are several advantages to using IMUs as a motion capture system since they are compact, light, resistant to environmental interference and easy to wear. These devices are robust and several successful examples of IMU measurement of upper limb movements have been described [3,16,17]

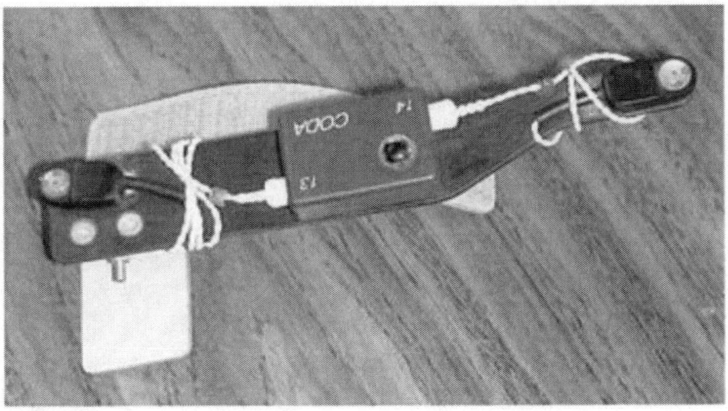

Figure 2. Active markers used for the kinematic gait analysis.

3. Central Cord Syndrome

The Central Ccord syndrome (CCS) is considered the most common incomplete SCI syndrome with a reported incidence varying from 15.7% to 25% [6]. Typically, the CCS results from a hyperextension of the cervical spinal cord and was first described by Schneider as a condition that is associated with sacral sparing and it is characterized by motor weakness that affects more the upper extremities than the lower limbs [18]. The prognosis for a functional recovery of CCS has been suggested to be rather favourable, depending of the age of the subject. Independent ambulation was achieved in 87-97% in younger patients compared to 31-41% in patients older than 50 years at the time of injury [19].

There are few studies on this type of population, and this could be due the lack of consensus regarding the diagnostic criteria. However, a recent review concluded established that the existence of a difference of at least 10 points between the motor index of upper and lower limbs served as a good diagnostic criterion for CCS [20].

Patients with CCS, with a lower speed than controls, show shorter strides and lower cadence, while the double support phase is longer. It has been reported that decreasing gait speed might be useful to prevent a fall when gait is perturbed [14]. These features have also been observed in different neurological diseases such as patients with spastic paraplegia, cervical myelopathy or Duchenne's muscular dystrophy [21-23].

It is well-known that walking speed conditions the kinematic gait variables. It is advised that control group should walk in this kind of studies with similar speed as the studied group. Nevertheless, it must be also bear in mind that this may introduce a certain bias in the data from the control group since walking slowly may modify their normal gait.

The biomechanical analysis of CCS patients reveals also a series of kinematic alterations such as a smaller range of motion of the knee and ankle in the sagittal plane and greater abduction of the hip at the initial contact and during the oscillation phase [24].

The normal peak of plantar flexion of the ankle is decreased in patients with CCS as occurs in other neurological disorders, contributing to the reduced walking speed [25].

The range of knee movement is decreased in the sagittal plane, whereby the knee tend to be more flexed during the support phase but less flexed during the oscillation phase. The limited knee flexion during the oscillation phase has been explained by the antagonistic action of the Rectus femoris muscle and of the Vastus lateralis, leading to the recommendation that strategies are adopted to stretch these muscles or other such adaptations of clinical treatments to improve these patients' capacity to walk [2]. This reduced range of knee flexion has been observed also in patients with paraplegic-spastic gait of diverse aetiology, in which this limitation was proposed to be correlated with the degree of spasticity [4]. Accordingly, this alteration also might be due to a specific loss muscle control [26].

Contrary to expected, the reduced joint movement of the knee and ankle in the sagittal plane has not been associated with an important reduction in the hip as seen in patients with other pathologies. This could be due to that the assessed patients had a mild form of the pathology.

4. Brown-Sequard Syndrome

The Brown Sequard syndrome (BSS) is caused by an injury that affects just one side of the spinal cord. True hemi-sections lesions of the spinal cord are rare, while partial lesions due to penetrating wounds (e.g.: gunshot wounds or knife penetrations) are more common. The clinical features are loss of motor function, proprioception, vibration, and light touch in ipsilateral side of the injury, and a loss of pain, temperature, and crude touch sensations in the contralateral one.

BrownSequard [18 60] was the first to relate these asymmetric clinical deficits to unilateral cord injury by noting similar asymmetric motor and sensory deficits in animals subjected to spinal cord hemi-section.

But such a pure form of BSS rarely occurs thus, “Brown-Sequard-plus syndrome” (BSPS) is now considered as a derivative of this state. BSPS encompasses SCI subjects suffering from asymmetric paresis combined with relatively pronounced analgesia on the less paretic side [20]. BSS is thought to have a better outcome than CCS, although the prognosis in both cases is quite good. [27,28] Between 87-97% of CCS patients aged under 50 years manage to walk independently while 75-90% of BSS patients can walk independently after rehabilitation [29].

The BSS has some characteristic biomechanics alterations related to its asymmetry [30]. The spasticity of plantarflexors in the affected side produces a maximum ankle dorsiflexion lower than controls. It is specially determinant with the tendency of the foot to be plantarflexed at the initial contact (Figure 4).

During the rest of the stance phase this tendency continues, appearing decreased ankle dorsiflexion during mid-stance and plantar flexion at toe off is increased.

Knee kinematics at sagital plane are disturbed with a significant reduction in the knee range of motion. That is shown in an increased flexion at initial contact but it is very relevant in the reduction of the maximum of flexion at toe-off. The could be a result of the effect of spasticity or, most probable, secondary to the decreased ankle dorsiflexion.

The knee shows also a decreased flexion at late stance as seen in patients with post stroke hemiparesis [31], but not enough to cause knee hyperextension..

The kinematic results for the hip and pelvis indicate that they also have greater internal rotation and that pelvic tilt ROM is increased.

Despite considering the BSS as a fundamentally unilateral abnormality, alterations in Motor Score can also be found in the less affected side.

The kinetic analysis shows mild disturbances in ground reaction forces (Figure 3). It has been described decreases in maximum antero-posterior force at loading response and maximum vertical force at terminal stance.

The moments at the ankle are related to tendency to plantarflexion (Figure 5), showing lesser maximums dorsiflexor moment. The knee has more extensor moment at the mid and late stance due to the extension at that point. The maximum hip flexor moment and the maximum hip abduction moment are also greater than in the gait of controls subjects. [30]

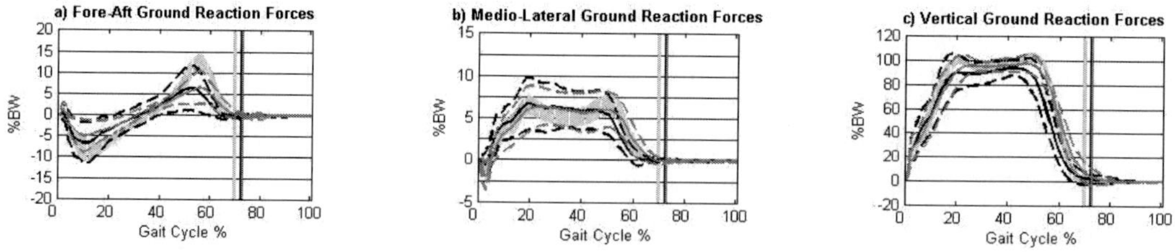

Figure 3. Kinetic analysis. Ground reaction force (GRF) curves: Fore-aft (Antero-posterior) (a), medio-lateral (b), and vertical (c). Mean (continuous thick line) and standard deviation (dashed line) for the BSS group or the more affected side (black), and of the CCS (grey). Control group walking at a slow gait speed (grey band including mean and standard deviation). The thick vertical line indicates toe-off. [30]

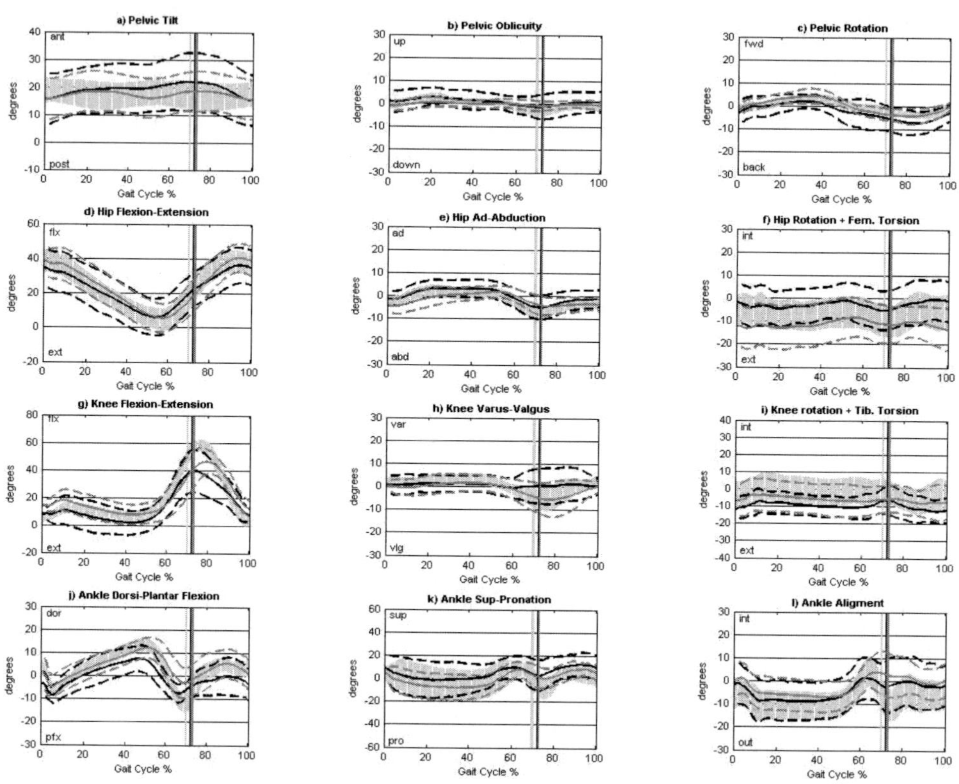

Figure 4. Kinematic curves in the sagittal (a,d,g,j), frontal (b,e,h,k) and horizontal (c,f,i) planes for the pelvis (a,b,c), hip (d,e,f), knee (g,h,i) and ankle (j,k). Mean (continuous thick line) and standard deviation (dashed line) for the BSS group or the more affected side (black), and of the CCS (grey). Control group walking at a slow gait speed (grey band including mean and standard deviation).The thick vertical line indicates toe-off [30].

5. Comparison

The comparison between these two Incomplete SCI syndromes can bring a better understanding of their particularities. The results of a pilot study with the aim to highlight the use of kinematic and kinetic analysis as an adjunct to assessment of individual patients with

Central Cord Syndrome and hemisection or Brown Séquard syndrome and to discuss their possible consequences for clinical management are presented in this chapter [30].

The analysis of temporal-spatial parameters of both groups reflected a cautious, abnormal gait, focused on balance and stability, but they also display a series of kinematic alterations as have been described independently.

Ankle kinematics seems to be the most remarkable difference between the two syndromes. The most remarkable differences found in the knee and ankle sagittal plane seems to be due to ankle position at initial contact. At initial contact the ankle was in a dorsiflexed position for CCS patients and in slight plantar-flexion for BSS. Ankle dorsiflexion values were decreased in BSS and plantar flexion value at toe off were increased in comparison to the CCS group.

The differences in ankle kinematics had different consequences at the knee for each of these types of patients. The increased knee flexion angle at initial contact in CCS could be related to the increased ankle dorsiflexion. The increased ankle plantarflexion could explain the decreased minimum knee flexion in BSS during the mid stance. Although both groups had increased knee flexion at initial contact, the slight plantar-flexion position in BSS could be the reason for the decreased minimum knee flexion during mid stance phase in this group. The ankle dorsiflexion at initial contact in CCS group could facilitate the increased knee flexion compared to BSS. This may be due to the supportive function of the arms of subjects with CCS is lesser than that of BSS subjects. Therefore, to achieve the comparable level of walking, subjects with CCS may have a higher motor score and better control of lower extremities than subjects with BSS to compensate for the deficit.

The knee joint ROM in the sagittal plane is reduced in both group of patients, and it is more flexed during stance for CCS.

With respect to the hip, the internal rotation in BSS is greater than in CCS. Internal rotation values are not reached by the CCS patients at any point in the gait cycle (Figure 4). Maximum hip external rotation seems to be greater in CCS patients than in BSS patients, due to the internal rotation tendency of the BSS2 subjects.

Walking at a comparable slow speed, the kinematic results for the pelvis and hip indicated that pelvic tilt ROM was greater in BSS.

Despite of this kinematic alterations, not just ROM limitations, but also motor control disorders due to lack of specificity in muscle control, can be responsible for these findings.

When analyzing both groups CCS and BSS in relation to controls, common findings in patient groups were a decreased knee ROM in sagittal plane and a decreased maximum knee flexion along gait cycle. However, knee flexion at initial contact was increased in CCS and BSS, especially in CCS. This abnormality can be considered due to motor control disorders, since no passive ROM limitations have to be present, This probably implies co-contraction of the anterior and posterior muscles of the thigh and could be due to a lack of specificity in muscle control, as has been reported [26]. Kinetic findings are shown in Figure 5.

Several limitations of this conclusions need to be discussed. First, the small number of patients included in the studies limits the validation of the statements. Even having similar types of lesions, there were a lot of heterogeneity in the groups of patients regarding gender, age, weight, time since injury and functional ability. These characteristics contributed to their effects of the injury onto the subjects and con make them walk differently and require different levels of support from walking devices. The different types of walking device also attributed crucial effects on gait patterns of the subject. This wide range of spinal lesion characteristics make investigations difficult due to its large standard deviations, limiting the ability to detect statistical significance.

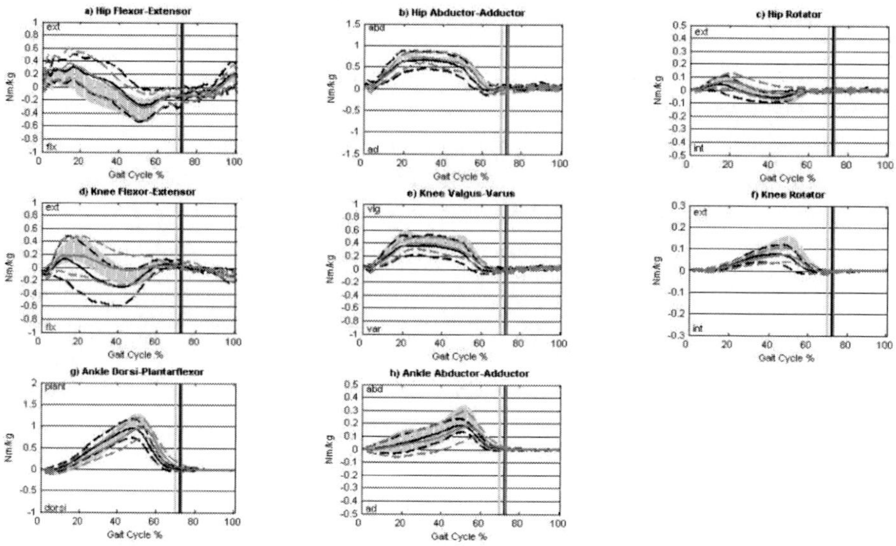

Figure 5. Kinetic analysis. Joint moment curves in the sagittal (a,d,g), frontal (b,e,h) and horizontal (c,f) planes of the hip (a,b,c), knee (d,e,f) and ankle (g,h). Mean (continuous thick line) and standard deviation (dashed line) for the BSS group or the more affected side (black), and of the CCS (grey). Control group walking at a slow gait speed (grey band including mean and standard deviation). The thick vertical line indicates toe-off [30].

Although the results shown on this chapter must be confirmed in a larger sample, these preliminary data could still be of interest because it is not usual to collect a large sample size of these two syndromes and, to date, we have not found comparative studies on biomechanical gait characteristics in CCS and BSS patients. Another possible limitation is that the CG subjects were required to walk at a slower speed so that the analysis of kinetic and kinematics data could be compared with the pathological population.

It is important to consider that gait speed influences certain kinetic and kinematics variables [32]. Recordings were obtained at a similar speed from all the groups to compare these variables. This consideration could introduce a certain bias, especially in the CG, since the subjects may modify their usual gait pattern in some way in order to walk at a slower speed. But this was the best way to compare CG findings with patients groups.

The small number of participants in the groups prevented the specific analysis of the contribution of the technical aids to the gait characteristics. In future research it should be considered to analyze changes after guiding task-oriented gait treatment by means of a biomechanical gait analysis in order to support a clinical application of gait biomechanical analysis. [11]

6. Treatment

Restoring and improving the capacity to walk is one of the main goals in CCS and BSS rehabilitation programs. In the past, the treatment of this group of patients used to consist on teaching them compensatory gait strategies. The gait analysis and the description of abnormal gait patterns produced by different neuro-motor disorders allows the clinics to develop new approaches for treatment and the definition of specific rehabilitation programs.

Due to the limited number of subjects analyzed and thus the few statistical data provided so far, it is very difficult to provide very specific treatment recommendations for a whole group. So, up to now, the treatments should be tailored considering the individual clinical state of each patient.

From a clinical point of view, the data obtained suggest that in patients with CCS, we should preferentially work on lengthening the hamstrings muscles and on muscle coordination to try to reduce the knee flexion at initial contact, and not only on strengthening the muscles. Indeed, while some studies indicate that an increase in strength in the lower limbs is related with an improvement in gait [33], others consider that this is not always the case [14].

From a clinical point of view, the reduction of knee flexion at initial contact in patients with CCS is suggested. This strategy could involve lengthening the knee flexor muscles and working on muscle coordination, rather than simply strengthening the muscles. Stretching the anterior Rectus femoris and the Vastus lateralis to help increase knee flexion during the oscillation phase and in general, to improve the range of knee mobility in the sagittal plane [2].

The treatment in BSS patients should focus on decreasing calf muscles hypertonicity and hyperactive stretch reflexes to increased plantar-flexion at initial contact. Different treatment options such as physiotherapy (calf muscles lengthening), orthoses (AFO) or pharmacological local treatment (Botulinum toxin) can be proposed to decrease this abnormality.

The last therapeutic options for walking disorders in incomplete SCI syndromes such as functional electrical stimulation (FES), botulinum toxin type A [34] or splinting have led to a improvement of treatments. The combination of the biomechanical gait analysis with these therapeutic possibilities allows better functional outcomes for patients These different approaches show the importance of gait analysis as a tool to make therapeutic decisions, such as the prescription of orthoses or a surgical procedure, and to evaluate the patient during treatment or after surgical intervention. It is also important to make a control of the evolution of the patient after therapeutic decisions. [30]

Conclusion

The treatment of gait disorders in incomplete spinal cord syndromes has been improved with specific therapies, which are able to return gait capacity or to correct abnormal gait patterns. The combination of the biomechanical gait analysis with these therapeutic possibilities allows better functional outcomes for patients. These different approaches show the importance of the individualization of the treatment based on gait analysis as a tool to make therapeutic decisions.

From a clinical point of view, the objective description of the gait cycle of CCS and BSS patients is interesting not only to establish their possible differences but also, to design specific rehabilitation treatments. The most relevant findings were in the knee and ankle sagittal plane. The main difference between both groups of patients lies at the initial contact, and although knee flexion increases in both cases, the ankle is in slight plantar-flexion in the BSS group and in slight dorsiflexion in the CCS group. In any case, treatment must aim to achieve a more secure gait with better motor control and better ankle positioning at initial contact.

References

[1] Go BK, De Vivo MJ, Richards JS. The epidemiology of spinal cord injury. In: Stover SL, DeLisa JA, Whiteneck GG, editors. Spinal cord injury: clinical outcomes from the model systems Gaithersburg (MD): *Aspen Publishing* 1995:21-55.

[2] Ditunno J, Scivoletto G. Clinical relevance of gait research applied to clinical trials in spinal cord injury. *Brain Res Bull* 2009 Jan 15;78(1):35-42.

[3] Requejo PS, Wahl DP, Bontrager EL, Newsam CJ, Gronley JK, Mulroy SJ, et al., Upper extremity kinetics during Lofstrand crutch-assisted gait. *Med Eng Phys* 2005 Jan;27(1):19-29.

[4] Krawetz P, Nance P. Gait analysis of spinal cord injured subjects: effects of injury level and spasticity. *Arch Phys Med Rehabil* 1996 Jul;77(7):635-638.

[5] Burns SP, Golding DG, Rolle W,Jr, Graziani V, Ditunno J,Jr. Recovery of ambulation in motor-incomplete tetraplegia. *Arch Phys Med Rehabil* 1997;Nov;78(11):1169-1172.

[6] McKinley W, Santos K, Meade M, Brooke K. Incidence and outcomes of spinal cord injury clinical syndromes. *J Spinal Cord Med* 2007;30(3):215-224.

[7] Nash B, Roller JM, Parker MG. The effects of tone-reducing orthotics on walking of an individual after incomplete spinal cord injury. *J Neurol Phys Ther* 2008 Mar;32(1):39-47.

[8] Fox EJ, Tester NJ, Phadke CP, Nair PM, Senesac CR, Howland DR, et al., Ongoing walking recovery 2 years after locomotor training in a child with severe incomplete spinal cord injury. *Phys Ther* 2010 May;90(5):793-802.

[9] Behrman AL, Nair PM, Bowden MG, Dauser RC, Herget BR, Martin JB, et al., Locomotor training restores walking in a nonambulatory child with chronic, severe, incomplete cervical spinal cord injury. *Phys Ther* 2008 May;88(5):580-590.

[10] Smith PA, Hassani S, Reiners K, Vogel LC, Harris GF. Gait analysis in children and adolescents with spinal cord injuries. *J Spinal Cord Med* 2004;27 Suppl 1:S44-9.

[11] Gil-Agudo A, Perez-Rizo E, Del Ama-Espinosa A, Crespo-Ruiz B, Perez-Nombela S, Sanchez-Ramos A. Comparative biomechanical gait analysis of patients with central cord syndrome walking with one crutch and two crutches. *Clin Biomech* (Bristol, Avon) 2009 Aug;24(7):551-557.

[12] Patrick JH. Case for gait analysis as part of the management of incomplete spinal cord injury. *Spinal Cord* 2003 Sep;41(9):479-482.

[13] Fritz SL, Merlo-Rains AM, Rivers ED, Peters DM, Goodman A, Watson ET, et al., An intensive intervention for improving gait, balance, and mobility in individuals with chronic incomplete spinal cord injury: a pilot study of activity tolerance and benefits. *Arch Phys Med Rehabil* 2011 Nov;92(11):1776-1784.

[14] Wirz M, Zorner B, Rupp R, Dietz V. Outcome after incomplete spinal cord injury: central cord versus Brown-Sequard syndrome. *Spinal Cord* 2010 May;48(5):407-414.

[15] Whittle MW. Gait analysis: an introduction - 3rd edition. Butterworth-Heinemann, Linacre House, Jordan Hill ed. Oxford. United Kingdom; 2003.

[16] Slavens BA, Sturm PF, Harris GF. Upper extremity inverse dynamics model for crutch-assisted gait assessment. *J Biomech* 2010 Jul 20;43(10):2026-2031.

[17] Cutti AG, Giovanardi A, Rocchi L, Davalli A, Sacchetti R. Ambulatory measurement of shoulder and elbow kinematics through inertial and magnetic sensors. *Med Biol Eng Comput* 2008 Feb;46(2):169-178.

[18] Schneider RC, Cherry G, Pantek H. The syndrome of acute central cervical spinal cord injury; with special reference to the mechanisms involved in hyperextension injuries of cervical spine. *J Neurosurg* 1954;Nov;11(6):546-577.

[19] Newey ML, Sen PK, Fraser RD. The long-term outcome after central cord syndrome: a study of the natural history. *J Bone Joint Surg Br* 2000;Aug;82(6):851-855.

[20] Pouw MH, van de Meent H, van Middendorp JJ, Hirschfeld S, Thietje R, van Kampen A, et al., Relevance of the diagnosis traumatic cervical Brown-S,quard-plus syndrome: an analysis based on the neurological and functional recovery in a prospective cohort of 148 patients. *Spinal Cord* 2010;Jan 12.

[21] Malone A, Meldrum D, Bolger C. Gait impairment in cervical spondylotic myelopathy: comparison with age- and gender-matched healthy controls. *Eur Spine J* 2012 Jul 24.

[22] D'Angelo MG, Berti M, Piccinini L, Romei M, Guglieri M, Bonato S, et al., Gait pattern in Duchenne muscular dystrophy. *Gait Posture* 2009 Jan;29(1):36-41.

[23] Ganea R, Jeannet PY, Paraschiv-Ionescu A, Goemans NM, Piot C, Van den Hauwe M, et al., Gait assessment in children with duchenne muscular dystrophy during long-distance walking. *J Child Neurol* 2012 Jan;27(1):30-38.

[24] Gil-Agudo A, Perez-Nombela S, Forner-Cordero A, Perez-Rizo E, Crespo-Ruiz B, del Ama-Espinosa A. Gait kinematic analysis in patients with a mild form of central cord syndrome. *J Neuroeng Rehabil* 2011 Feb 2;8:7.

[25] Nadeau S, Gravel D, Arsenault AB, Bourbonnais D. Plantarflexor weakness as a limiting factor of gait speed in stroke subjects and the compensating role of hip flexors. *Clin Biomech* (Bristol, Avon) 1999 Feb;14(2):125-135.

[26] Alexeeva N, Broton JG, Suys S, Calancie B. Central cord syndrome of cervical spinal cord injury: widespread changes in muscle recruitment studied by voluntary contractions and transcranial magnetic stimulation. *Exp Neurol* 1997;Dec;148(2):399-406.

[27] Woosli RM YR. The clinical diagnosis of disorders of the spinal cord. In: Young RM WR, editor. *Diagnosis and Management Disorders of the Spinal Cord.:* Saunders: Philadelphia. p. 135--144.

[28] Aito S, D'Andrea M, Werhagen L, Farsetti L, Cappelli S, Bandini B, et al., Neurological and functional outcome in traumatic central cord syndrome. *Spinal Cord* 2007;Apr;45(4):292-297.

[29] Hayes KC, Hsieh JT, Wolfe DL, Potter PJ, Delaney GA. Classifying incomplete spinal cord injury syndromes: algorithms based on the International Standards for Neurological and Functional Classification of Spinal Cord Injury Patients. *Arch Phys Med Rehabil* 2000 May;81(5):644-652.

[30] Gil-Agudo A., Pérez-Nombela S., Pérez-Rizo E., del Ama-Espinosa A., Crespo-Ruiz B. Comparative biomechanical analysis of gait in patients with central cord and Brown-Sequard syndrome. *Disability and Rehabilitation.* Article in press.

[31] Kim CM, Eng JJ. Magnitude and pattern of 3D kinematic and kinetic gait profiles in persons with stroke: relationship to walking speed. *Gait Posture* 2004 Oct;20(2):140-146.

[32] Ochi F, Esquenazi A, Hirai B, Talaty M. Temporal-spatial features of gait after traumatic brain injury. *J Head Trauma Rehabil* 1999;14:105.--115.

[33] Kim CM, Eng JJ, Whittaker MW. Level walking and ambulatory capacity in persons with incomplete spinal cord injury: relationship with muscle strength. *Spinal Cord* 2004;Mar;42(3):156-162.

[34] Santamato A, Panza F, Ranieri M, Amoruso MT, Amoruso L, Frisardi V, et al., Effect of intrathecal baclofen, botulinum toxin type A and a rehabilitation programme on locomotor function after spinal cord injury: a case report. *J Rehabil Med* 2010 Oct;42(9):891-894.

In: Gait Biometrics
Editors: Li Li and Matthew Holmes

ISBN: 978-1-62948-381-8

Chapter VII

The Effects of Altered Lower Limb Mechanics as a Result of Amputation on Gait and Posture and the Implications for Falls

Natalie Vanicek*
Discipline of Exercise and Sport Science, Faculty of Health Sciences,
University of Sydney, Australia

Abstract

Following the amputation of the lower limb, a prosthesis is typically prescribed to provide the structural support needed for walking. Inherently, lower limb amputees have modified locomotor systems due to the loss of muscle function, diminished joint movement and modified feedback from muscles and receptors. Yet, they can relearn to walk proficiently and comfortably within their new biomechanical constraints. However, there are many factors that predispose lower limb amputees to falling, and they experience more falls per year than age-matched, able-bodied adults. Previous research has reported that 52-58% of individuals with a major lower limb amputation experience at least one fall within a 12-month period. Falls are a major healthcare problem and have serious physical, psychological and socioeconomic consequences. Gait disorders that are related to reduced lower limb strength and joint range of motion, muscle tightness and postural instability are associated with an increased risk of falling in older people. Therefore, lower limb amputees and older adults exhibit similar characteristics that predispose them to falling. Whilst falls are a serious problem for our ageing population and lower limb amputees, effective and targeted rehabilitation and exercise programmes have the potential to improve locomotor performance, reduce falls and improve general health and quality-of-life. This chapter focuses on the risk factors for falls in lower limb amputees and describes some of the biomechanical characteristics of lower limb amputee

* Corresponding author: Dr Natalie Vanicek, Discipline of Exercise and Sport Science, Faculty of Health Sciences, University of Sydney, Australia. E-mail: natalie.vanicek@sydney.edu.au.

gait. A general overview about post-operative procedures and prosthetic gait training is described. The epidemiology of falls is reviewed and the literature focusing on the gait differences between fallers and non-fallers is discussed critically. Common activities of daily living such as level, inclined and stair walking, turning during walking and loss of balance in dynamic situations are often precursors to a fall. This chapter examines the functional demands of such activities and makes evidence-based recommendations for targeted exercises for lower limb amputees. This chapter concludes with an overview of the relative popularity of game console use as a novel training tool in amputee rehabilitation and how this may impact physiotherapy and falls prevention programmes in patients with a high falls risk.

Introduction

An amputation is the surgical removal of a limb and occurs primarily as a result of disease, trauma, infection or congenital deformities. In the lower limbs, an amputation below the knee (transtibial) is more frequent than above the knee (transfemoral). Whenever possible, the surgeon will try to preserve anatomical joints and soft tissue by performing the amputation at the most distal level possible, as the energy requirements of walking with an amputation at the transfemoral level are generally higher than with a transtibial amputation. While a prosthesis may provide the necessary structural support to ambulate, it lacks the ability to detect changes in limb position, muscle strength and joint flexibility. The majority of lower limb amputees have had their limb removed due to vascular reasons related to peripheral arterial disease or diabetes and are over the age of 60 years. Consequently, many lower limb amputees have confounding health problems and are at a particularly high risk of falling.

Amputee Statistics

The vast majority of lower limb amputations occur at the transtibial level. Transtibial amputations comprised 53% of all lower limb amputations performed in the UK in 2006-07, while 39% were at the transfemoral level. The remaining 8% of amputations included hip and knee disarticulations, partial foot amputation or bilateral amputation (NASDAB, 2009). Contrary to the UK, the US does not have a national surveillance system for monitoring limb loss, which makes it difficult to present and compare data with other developed nations, but it is estimated that there are approximately 1.6-1.7 million people living with an amputation in America (Ziegler-Graham et al., 2008). Predictions indicate that the number of people with an amputation will more than double to 3.6 million by the year 2050, confirming that amputation is a growing health concern in developed nations. In the US, dysvascular-related amputations accounted for 82% of limb loss hospital discharges (NLLIC, 2008). The patterns related to gender and age are very similar across countries. In the US, 42% of persons living with an amputation were 65 years or over and 65% of these people were men. In the UK, more than two thirds of all amputations are performed in men, especially aged over 65 years (Figure 1). Conversely, more women than men are referred from the oldest age category of > 75 years of age, and women have an almost equal likelihood of experiencing a transtibial or transfemoral amputation. The overarching cause of lower limb amputation in the UK and the US is related to dysvascularity. This term encompasses the two main contributors diabetes mellitus and non-diabetic arteriosclerosis, but also includes causes such as embolisms, rheumatoid arthritis, autoimmune disease and venous disease, among others.

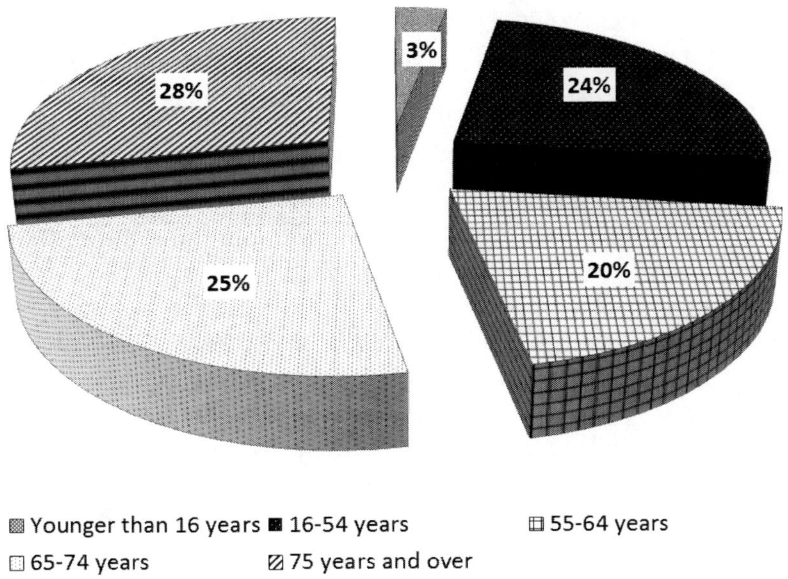

Adapted from NASDAB, 2009.

Figure 1a. The percentage of new referrals to lower limb prosthetic services in the UK according to age group in 2006/07.

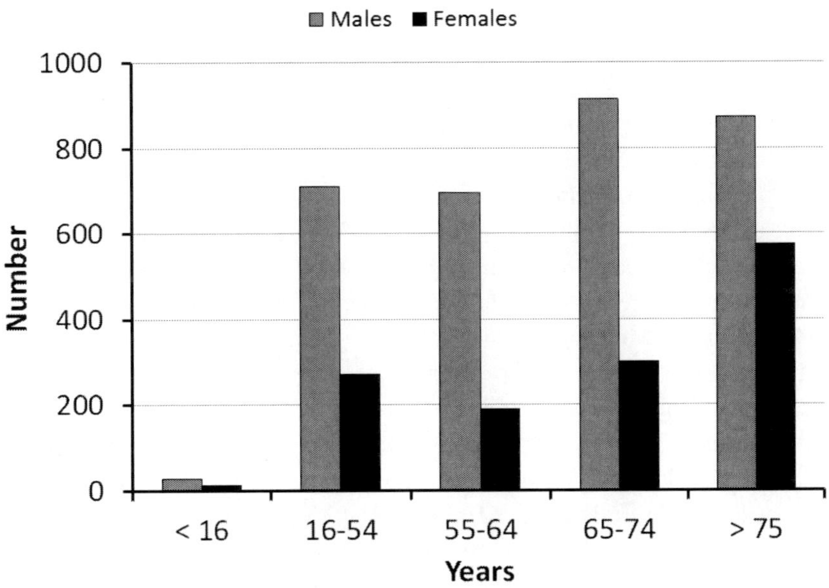

Adapted from NASDAB, 2005/6.

Figure 1b. Number of new referrals to prosthetic limb centres in the UK according to age and gender.

An amputation caused by trauma, such as mechanical (including motor vehicle accidents) or other electrical or chemical incidents, is less common. Other causes included cancer, infection, neurological disorders, neoplasia or remain unreported. While the number of amputations has remained relatively constant in the UK, the proportion of amputations due to dysvascularity has risen over a ten-year period ending in 2007 (NASDAB, 2009). Such trends are expected to continue as vascular disease and diabetes are becoming global epidemics.

Conversely, in developing countries and those that have seen recent warfare, amputations are often related to trauma, inadequately treated fractures or infections, and motor vehicle accidents.

Amputee Gait Characteristics

Lower limb amputees are able to regain their mobility with a prosthesis by learning to adapt to a new set of biomechanical constraints. However, the gait patterns of transtibial and transfemoral amputees differ somewhat from able-bodied gait. Some of the most basic differences are that lower limb amputees walk more slowly and yet expend more energy per distance walked. Even though the mass of the residual limb and prosthesis is lighter than the unaffected side, the metabolic requirements increase as the level of amputation moves proximally and the length of the residual limb decreases. Unilateral amputees also exhibit asymmetrical gait patterns, by taking longer steps with the amputated limb and spending less time supporting their body weight on the amputated side. These are considered compensatory mechanisms to protect the residual limb. One might expect that if the mass and inertial properties of the prosthesis were matched to the unaffected limb, gait would become more symmetrical and energy costs would decrease. However, a study by Mattes and colleagues (2000) found that matching the mechanical characteristics had the opposite effect by significantly increasing energy costs and contributing to a more asymmetrical gait pattern in transtibial amputees. The kinematic (joint motion) patterns will vary according to the mechanical capabilities of the prosthesis, ankle-foot complex and mechanical knee joint. Some of the most noticeable differences are observed at the ankle joint, as the prosthetic ankle does not facilitate active plantarflexion. In transtibial amputees, the biological knee often shows reduced flexion during the loading response, and in the case of transfemoral amputees, may actually be locked in extension during stance, such as when using a mechanically-passive prosthetic knee joint. By maintaining the knee more extended, stability is enhanced as the ground reaction force vector passes through or close to the knee joint, thereby keeping the knee joint moment close to zero. Such gait deviations help to reduce the (eccentric) muscular demands of the weakened knee extensor muscles. Conversely, transfemoral amputees who use a microprocessor-controlled knee joint have enhanced knee stability and demonstrate shock absorption, by flexing the prosthetic knee during the loading response. A microprocessor-knee joint, such as the C-leg, Genium knee (both OttoBock) and Intelligent Prosthesis (Endolite Blatchford), has variable-damping, which assists with flexion during swing (Johansson et al., 2005). However, these more advanced prosthetic devices are costly, and generally prescribed for younger or more active amputees. Kinematic profiles for the hip are more stereotypical than for the ankle or knee, although lower limb amputees may show less hip extension. This could be related to forward trunk flexion, such as when walking with an assistive device, slower walking speed and reduced step length, or tightness in the hip flexor muscles (such as the iliopsoas group, rectus femoris, sartorius) or a combination of any of these factors. During level gait, lower limb amputees exhibit reduced kinetics by generating smaller propulsive forces and reducing the rate at which the amputated limb is loaded. Joint moment and power profiles are also modified, and compensations for the lack of power generation by the ankle plantarflexor muscles occur primarily at the hip joint on the amputated side.

Knee joint moments are low and contribute little to the overall support moment. Muscle activity profiles, as measured using surface EMG, demonstrate that lower limb amputees compensate for an amputation with muscles proximal to the site of amputation. The hip extensors (gluteus maximus) are hyperactive and for a longer period of time to facilitate push-off in the absence of the power generating ankle plantarflexors (triceps surae). Moreover, surface EMG profiles reveal opposing knee extensor and flexor muscles are active simultaneously (Winter and Sienko, 1988). Termed co-contraction, muscle activity from the agonists and antagonists helps to stabilise the knee joint. This is an adaptation that enhances prosthetic stability during stance in transtibial amputees. These aforementioned musculoskeletal alterations help to ensure lower limb amputees can manoeuvre different types of activities of daily living safely despite their biomechanical constraints.

Post-Operative Procedures

Post-operative procedures can be difficult to generalise as they may depend on a multitude of factors including the patient's medical condition, existing health comorbidities, pre-operative mobility, wound healing and swelling, and psychological factors. For these reasons, and when possible, it is appropriate that a patient's rehabilitation begins pre-operatively to facilitate the best possible adjustment to their new physical constraints. Esquenazi (2004) described nine phases of rehabilitation: (1) pre-operative, (2) surgical, (3) acute post-surgical (4) pre-prosthetic, (5) prosthetic prescription, (6) prosthetic training, (7) community integration, (8) vocational rehabilitation and (9) follow-up. These phases will vary according to the patient's level of expected mobility and functional level of classification.

Prior to the amputation and when feasible, the surgeon and patient will discuss post-operative goals and expectations and the patient's physical status is established. During the surgical process, the surgeon attempts to preserve the longest residual length as possible as this facilitates prosthetic fitting and functional mobility later on. Following surgery, the patient will stay in an acute hospital setting for approximately 2-5 days. Compression sleeves are fitted around the residual limb to help reduce oedema, promote limb shaping and wound healing. Pain is controlled with appropriate medication although treating phantom pain, which is pain experienced in the missing limb, is more complex. Pre-prosthetic rehabilitation may begin as early as 1-2 days following surgery. Due to immobilisation following surgery, muscle contractures are common occurrences and patients typically lose range of motion at the hip and knee. Therefore maintaining full knee extension and at least 20° extension at the hip are important pre-prosthetic goals achieved through exercise. Strengthening the critical muscles in preparation for ambulation with a prosthesis should also begin as soon as possible. Compression sleeves are used daily, but the skin should be left to air for at least 1 hour/day and frequently checked for possible irritations and/or signs of infection. The residual limb is monitored and its shape is measured before prosthetic fitting and prescription can take place. For transtibial amputees, the residuum should be cylindrical, while for transfemoral amputees it should be more conical. Casting for a preliminary prosthesis can be done between 3-5 weeks and partial weight-bearing gait training begins. Once the patient is fitted with a prosthesis, gait and other daily activities can be practised first within parallel bars and then progressing to elbow crutches or another assistive device.

This process is described in more detail in the next section. A patient may wear their prosthesis for only 1-2 hours/day initially, and progress according to their individual ability. It is very difficult to anticipate the duration of prosthetic rehabilitation, and increasingly so as the majority of amputees are also older, dysvascular patients. Transtibial amputees may attend sessions for 4-6 weeks or longer before being discharged from physiotherapy; transfemoral amputees require more time (6-12 weeks) and bilateral amputees usually need at least 3 months before becoming independent ambulators. Patients might attend a minimum of 3 sessions/week but are also expected to complete home-based exercises. A multitude or physical or mental factors can affect a patient's progress at any time during the course of rehabilitation. Some patients might require more formal training according to their requirements in the home or workplace; physical and occupational therapists in the multidisciplinary team can assist with such transitions. Although a patient is formally discharged from active physiotherapy or gait rehabilitation, amputees attend frequent follow-up sessions to ensure correct prosthetic fitting, undergo general medical assessments, and receive psychological support when and if needed.

Prosthetic Gait Training

There are many advantages of early gait re-education, such as reducing oedema in the residual limb, improving prosthetic and (partial) weight-bearing confidence, resuming a standing posture, and accelerating the time for prosthetic casting (Scott et al., 2000). Early walking aids (EWAs) are used solely in a hospital or rehabilitation setting and are not intended for community ambulation. In the UK, the Chartered Society of Physiotherapists guidelines advocate the use of EWAs as an important aspect of the rehabilitation process for lower limb amputees. An EWA is a generic prosthetic device and consists of an inflatable sleeve placed around the residual limb, which is then positioned within a rigid frame with a rocker foot. Patients are able to practice walking even before their residuum is ready to be cast for a prescribed prosthesis. There are currently two types of EWAs commonly used with transtibial amputees by physiotherapists in the UK: the Pneumatic Post-Amputation Mobility (PPAM) Aid that locks the biological knee in extension (stiff-knee gait) and the Amputee Mobility Aid (AMA) that incorporates a hinge joint, thus allowing the biological knee to flex and extend and facilitating a more natural gait pattern (Figure 2). While it might be hypothesised that the AMA should ease the transition to the definitive, prescribed prosthesis and encourage a better walking style, one study that investigated the gait differences between these two EWAs found no differences. Barnett and colleagues (2009) collected kinematic data from transtibial amputees during specific time points in their physiotherapy treatment and compared the gait characteristics of patients who initially walked with the PPAM aid or the AMA.

Their findings did not support the hypothesis that the AMA would show greater advantages during gait retraining. Moreover, neither EWA showed a clinical benefit in terms of duration of rehabilitation, number of visits or overall effect on health-related quality of life factors (Mazari et al., 2011). These studies have demonstrated that the most significant gait adaptations occur during the prosthetic training phase with a functional prosthesis, and suggest that early weight-bearing with an EWA is more important than practising knee control during the pre-prosthetic phase.

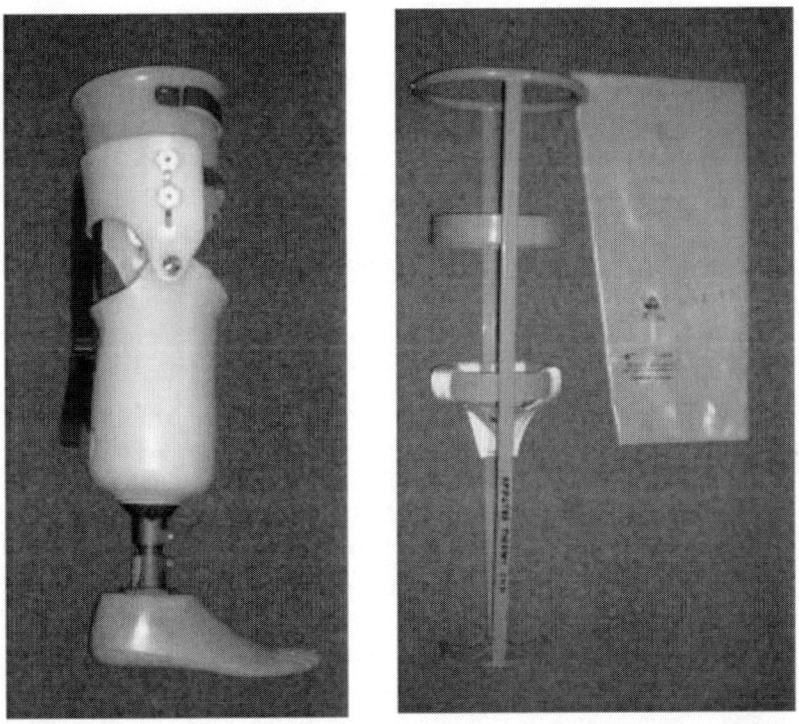

a b

Image used with permission (Ortho Europe Ltd, Ability House, Nuffield Way, Abingdon, Oxon, OX14 1RL, UK) (+44 (0)1235 552 895, info@ortho-europe.com, www.ortho-europe.com).

Figure 2. (A) The Amputee Mobility Aid (AMA) and (b) Pneumatic Post Amputation Mobility (PPAM) Aid used as early walking aids (EWAs) in the pre-prosthetic phase of rehabilitation in transtibial amputees.

Another possible reason for the use of an EWA is to screen patients for prosthetic prescription and compliance. Although the PPAM Aid may be used with transfemoral amputees, the Femurett is a generic, adjustable training prosthesis used uniquely for those with a Gritti-Stokes (through the knee) or above the knee amputation. The Femurett consists of a uniaxial 'ankle-foot', a tubular pylon, single axis knee joint and an adjustable thigh section, with shoulder strap for suspension.

Compared to the PPAM Aid, the rigid socket of the Femurett showed better stability and also permitted better gait evaluation, as the residuum was not contained within an inflatable bag (Parry and Morrison, 1989). The Femurett did not speed up the time spent in gait re-education, but it did create a more realistic prosthetic experience than the PPAM Aid, by incorporating a single-axis knee joint and foot.

Falls in Amputees

A fall has been described as "an event which resulted in the person coming to rest inadvertently on the ground or other level, other than as a consequence of lost consciousness, a violent blow, stroke or epileptic seizure" (Askham, 1990). Falls and their repercussions have become a serious health concern among able-bodied older adults.

It has been reported that 20-40% of adults aged over 65 years report one fall every year, and this number increases twofold for older adults living in institutions (Tinetti et al., 1986; American Geriatrics Society, 2001). However, among lower limb amputees, falls are even more frequent occurrences. A previous study found that 52% of lower limb amputees experience a fall every year, and amongst those fallers, that 75% were recurrent fallers, having fallen twice or more. These statistics demonstrate that amputees fall considerably more than their age-matched, able-bodied counterparts (Miller et al., 2001). These values may even be under-representations as many falls go unreported. The study by Miller et al., (2001) identified risk factors for falling, which included having a transfemoral amputation, being a more recent amputee (i.e., having less prosthetic practice), experiencing joint and back pain, and having problems with the prosthesis. Contrary to previous findings reporting vascular amputees exhibit greater postural sway than traumatic amputees (Hermodsson et al., 1994), having an amputation related to vascular disease was not an independent risk factor for falling. On the other hand, having walked with a prosthesis for at least 4 years served as a protective factor against falls. Falls are often accompanied by a fear of falling, which in turn may lead to activity avoidance, reduced social activity and poorer quality of life. Miller et al., (2001) reported that concentrating on walking and having a falls history were risk factors for developing a fear of falling; conversely, being male and having good overall perceived health were deemed protective factors.

Much of the falls literature over the past ten years has focused on identifying age-associated biomechanical alterations (Kerrigan et al., 1998; DeVita and Hortobagyi, 2000) and differences between fallers and non-fallers in older (able-bodied) adults (Kerrigan et al., 2000). Compared to young adults, older individuals walk more slowly (Kerrigan et al., 1998) and redistribute the joint moments away from the ankle and knee by compensating at the hip (DeVita and Hortobagyi, 2000).

Moreover, previous (older) fallers tend to walk more slowly than (older) non-fallers and young adults (Kerrigan et al., 1998; Kerrigan et al., 2000). Kerrigan and colleagues (2000) reported that four specific kinetic parameters differentiated older fallers from non-fallers and were independent of speed. The fallers demonstrated a greater internal hip extensor moment during stance, but reduced internal hip flexor moment, internal knee extensor moment and knee power absorption, all in pre-swing.

These results were corroborated by DeVita and Hortobagyi (2000) in that older adults exhibit increased output by the hip extensor muscles. Overall, reduced joint mobility, particularly at the ankle and hip, ankle plantarflexor muscle weakness and hip flexor tightness are considered risk factors for falling in older adults.

Many of the aforementioned musculoskeletal risk factors for falls in older adults can also be generalised to lower limb amputees. For example, lower limb amputees typically have a slower preferred walking speed and asymmetrical gait patterns. The missing ankle plantarflexor muscles on the amputated side mean the amputated limb lacks the propulsive capabilities of the unaffected limb. The ankle joint is unable to plantarflex actively; the knee joint range of motion (in the case of transtibial amputees) may be limited by the prosthetic socket on the posterior side of the knee while in transfemoral amputees the often-prescribed mechanical knee is locked in extension during stance. Hip flexor contractures are common, and hip extension is limited. If an amputee walks with an assistive device, they are likely to exhibit forward trunk flexion, which is accompanied by anterior pelvic tilt. All of these gait characteristics are synonymous with high falls risk.

Surprisingly few studies have investigated the biomechanical characteristics that may differentiate amputee fallers from non-fallers, given that this topic has received so much attention among able-bodied older adults, and falls are a recognised problem among amputees. Vanicek and colleagues (2009a) examined whether transtibial amputee fallers exhibited reduced joint mobility, and altered joint moments and powers compared to amputee non-fallers. Contrary to their hypotheses, there were no differences in walking speed or joint mobility between the fallers and non-fallers. By examining the joint kinetics, it became evident the fallers had higher vertical ground reaction forces and loaded their amputated limb significantly more than the non-fallers. Amputee fallers showed less eccentric contraction by the ankle plantarflexor muscles on the unaffected side during stance. This could have an adverse affect on the ability to position the prosthetic foot safely on the ground as the amputated limb transitions from swing into stance. The fallers also displayed greater demands by the hip flexor muscles to decelerate the thigh in late stance on the unaffected side. Inadequate eccentric hip muscle strength during this vulnerable transition into prosthetic single support could be a risk factor for falls in amputee. The biomechanical differences revealed that the amputee non-fallers adopted a more cautious gait pattern, and particularly during the transition from double limb support to single support on the amputated side. These findings have important implications for gait re-training: targeted exercise interventions can be designed by identifying instances in the gait cycle when muscular demands are high.

Activities of Daily Living in Lower Limb Amputees and Differences in Fallers vs Non-Fallers

Turning

We do not walk purely in a straight line and most of our daily activities involve turning. It is important to recognise the biomechanical demands of various tasks and how these demands vary with altered lower limb mechanics. Glaister et al., (2007) explored the types of steps taken over 4 different courses: walking from the office to the car park (252 m); shopping for groceries in a small convenience store (82 m); walking from one office to another (47 m); and walking through a cafeteria (68 m). The percentage of steps that were categorised as turning steps were 8%, 35%, 45%, and 50%, respectively. In a study with able-bodied participants, Strike and Taylor (2009) demonstrated that participants alter approach foot placement and velocity in preparation for a 90° turn and that turning relies on the ability to modulate ground reaction force impulses in all three dimensions. Turning involves re-orienting the COM in a new direction without stopping. Most right-handed and right-footed (able-bodied) individuals prefer to use the right foot as the *stabiliser* and so turn to the left (left foot acts as the *mobiliser*) (Taylor et al., 2006). Turning away from the stance or stabilising limb (called a *step turn*) maintains the COM between the two limbs; conversely turning towards the stance limb (called a *pivot or crossover turn*) moves the COM outside the base of support. In unilateral lower limb amputees, turning bias may be less obvious. In a sample of transtibial amputees, no significant preference was found for turning away from the dominant foot and turning bias was not related with the side of amputation.

In fact 59% of transtibial amputees preferred turning towards their amputated side, which allowed the unaffected side to act as the stabiliser when initiating the turn (Taylor et al., 2006). In order to compensate for the missing musculature and joint ROM at the knee and/or ankle, and the decreased internal rotation moments of the amputated limb, amputees make adaptations with their unaffected limb by generating larger external hip rotation moments when it is the inside leg on a circular path.

This may serve as a protective strategy to reduce the joint moments on the amputated side (Segal et al., 2011). As turning gait is known to be more difficult for individuals with mobility impairments, including amputees, more research examining the relationship between falls incidence and the mechanical demands of turning is warranted.

Inclined Walking

Lower limb amputees encounter more demanding environments than just level walkways. For example, wheelchair-access to buildings is facilitated by the presence of ramps. Walking on an inclined surface with a slope of 5%, the maximum gradient advised by buildings regulations, presents a demanding task for lower limb amputees, as they require greater slip resistance than able-bodied individuals (Buczek et al., 1990).

Few studies have investigated the biomechanical differences between lower limb amputees and able-bodied individuals, and those that have showed somewhat conflicting findings. Vickers et al., (2008) measured lower limb gait parameters in transtibial amputees and reported that the amputees walked significantly slower (0.71 m/s) than the able-bodied participants (1.41 m/s) uphill and downhill. However, Vrieling et al., (2008a) found that transtibial (1.21 m/s) and transfemoral (1.01 m/s) amputees walked with similar speeds to able-bodied participants (1.34 m/s) during inclined walking. These differences may be related to cause of amputation, as the majority of participants in Vickers et al., (2008) experienced an amputation related to dysvascularity, not trauma.

During inclined walking, Vickers et al., (2008) demonstrated that the hip, knee and ankle profiles of the transtibial amputees had reduced ROM for both the unaffected and amputated limbs; neither limb moved into extension at the hip and neither limb showed knee flexion during the loading response.

Conversely, Vrieling et al., (2008a) revealed that trantibial amputees exhibited similar adjustment strategies to able-bodied participants by showing greater hip and knee flexion (at initial contact and in swing) during uphill walking, and greater knee flexion (in late stance and swing) during downhill walking on the amputated side. Transfemoral amputees were unable to make adaptations at the prosthetic knee joint and thus, compensated at the hip in such a way so as to shorten step length on the amputated side during downhill walking. Shorter step lengths effectively reduce the height difference between steps when walking on an incline.

The unaffected limb plays an important role in compensating for the mechanical deficits of the prosthesis, as evidenced by increased knee flexion during the loading response, which assists to lift the body during uphill walking. In lower limb amputees, the shorter amputated limb length facilitates lowering of the body during downhill walking; but during uphill walking, greater demands are placed on the unaffected limb to increase the amount of positive work it performs.

Despite the higher falls risk associated with inclined walking, no published studies to date have explored the biomechanical differences between amputee fallers and non-fallers walking up- or downhill and such research would provide valuable information for therapists involved in the design of exercise programmes for amputee rehabilitation.

Stair Walking

Negotiating stairs is more mechanically demanding compared to level walking (Beaulieu et al., 2008; McFadyen and Winter, 1988; Reeves et al., 2008a; Reeves et al., 2008b) and the threat of tripping is greater as step length and height are dictated by the dimensions of the stairs. In older individuals or individuals with impaired mobility, falls often occur on stairs (Talbot et al., 2005) and with severe consequences. Both stair ascent and descent rely on muscle power particularly from the ankle plantarflexors and knee extensors. In lower limb amputees, the ankle plantarflexors are absent and the knee musculature on the amputated side, particularly the extensors, is weakened or inactive because of an extended knee position. Stair walking places greater functional demands on the prosthesis (Aldridge et al., 2012; Schmalz et al., 2007) and moves the lower limb joints through a greater range of motion than level walking (Lin et al., 2005).

Several studies have previously investigated the biomechanics of stair walking patterns in transtibial amputees, during stair ascent and/or descent over several steps (Alimusaj et al., 2009; Aldridge et al., 2012; Powers and Boyd, 1997; Schmalz et al., 2007; Yack et al., 1999) and in transfemoral amputees (Bae et al., 2009; Hobara et al., 2012; Schmalz et al., 2007). However, few have conducted both kinematic and kinetic analyses in this population because of the need for an instrumented staircase. These studies reported that transtibial amputees walk slower than control subjects. During the pull-up phase of stair ascent, the relative contribution of the ankle joint and the soleus muscle is high (McFadyen and Winter, 1988). Joint kinematic profiles from the amputated side reveal severely limited ankle range of motion, with an ankle-foot complex maintained in dorsiflexion. Thus, on the amputated side, this work must be divided between the knee and hip musculature, or in the case of transfemoral amputees, by the hip primarily, with some help from the unaffected limb during the double support phase (Yack et al., 1999). Yack et al., (1999) reported greater reliance on the hip extensor muscles on the amputated side during the weight acceptance and pull-up phases. These authors emphasised the importance of hip muscle strengthening strategies to facilitate an alternate, or step-over-step, stair walking pattern in amputees (Yack et al., 1999).

At the knee joint, results have been more confounding. Alimusaj et al., (2009) and Powers and Boyd (1997) reported similar knee flexion profiles during stair ascent in amputees and able-bodied individuals but reduced flexion angles during descent in the amputee group (Figure 3); while Schmalz et al., (2007) reported reduced and delayed knee flexion in stance during descent. Such reported inconsistencies may be caused by the different dimensions of the staircases used in these studies. Other kinematic data showed increased hip flexion (Alimusaj et al., 2009; Powers and Boyd, 1997; Schmalz et al., 2007) and anterior pelvic tilt (Powers and Boyd, 1997) during both ascent and descent when compared to able-bodied.

Kinetic results show that amputees have reduced peak vertical and decelerating ground reaction forces on the amputated leg but higher vertical GRFs on the unaffected leg during both stair ascent and descent (Schmalz et al., 2007).

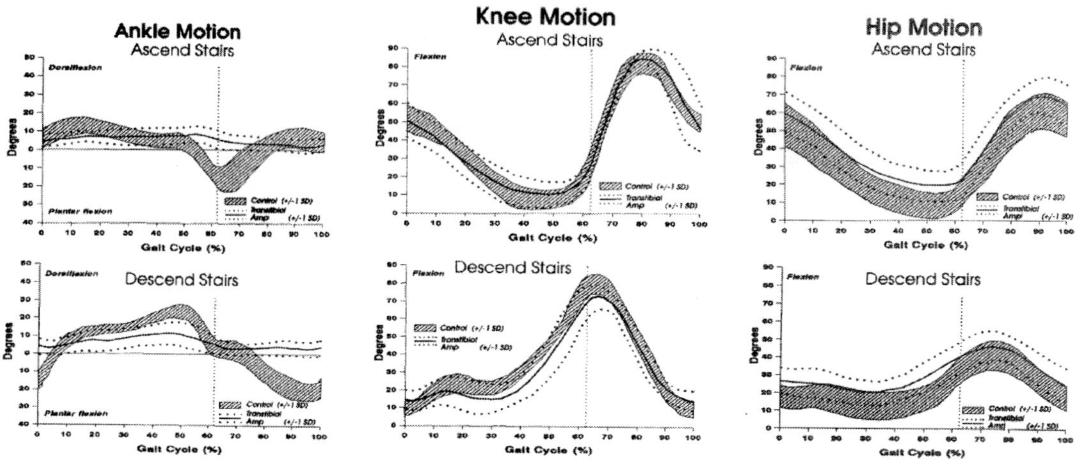

Figure 3. Mean joint motion (±1 standard deviation) of the ankle, knee and hip for controls (n = 14) and those with transtibial amputation (n = 10) during stair ascent and descent (Powers and Boyd, 1997).

Joint moment profiles have revealed a substantially smaller ankle joint moment during stair ascent. A reduced knee extensor moment on the amputated side maintains the knee in a more extended position (Alimusaj et al., 2009; Schmalz et al., 2007) and particularly so in transfemoral amputees (Hobara et al., 2012). Conversely, hip joint extensor moments on the amputated side are larger in magnitude compared to able-bodied individuals during stair ascent (Alimusaj et al., 2009; Schmalz et al., 2007; Vanicek et al., 2010). Previous authors have concluded that lower limb amputees employ a knee-extensor dominant strategy on the unaffected side (Yack et al., 1999) and a hip-extensor dominant strategy on the amputated side (Alimusaj et al., 2012) such that the hip extensors were the main source of power generation during stair ascent.

During stair descent, one of the most striking differences between amputees and able-bodied individuals is that amputees tend to make initial contact with the heel, or mid-foot whilst able-bodied land on the ball of the foot as it is plantarflexed (Alimusaj et al., 2009; Schmalz et al., 2007). This leads to very different ankle joint moment profiles. Transfemoral amputees use a distinct strategy by landing with the heel near the nose of the step; the foot can then roll over the edge. Prosthetic knee components that can generate flexion resistance, such as the C-leg microprocessor knee, are important features that facilitate this gait pattern (Schmalz et al., 2007). Upon landing, amputees attempt to position the centre of mass directly above the limb, thus reducing the lever arm of the ground reaction force vector and minimising the requirements of the knee extensors (Jones et al., 2006).

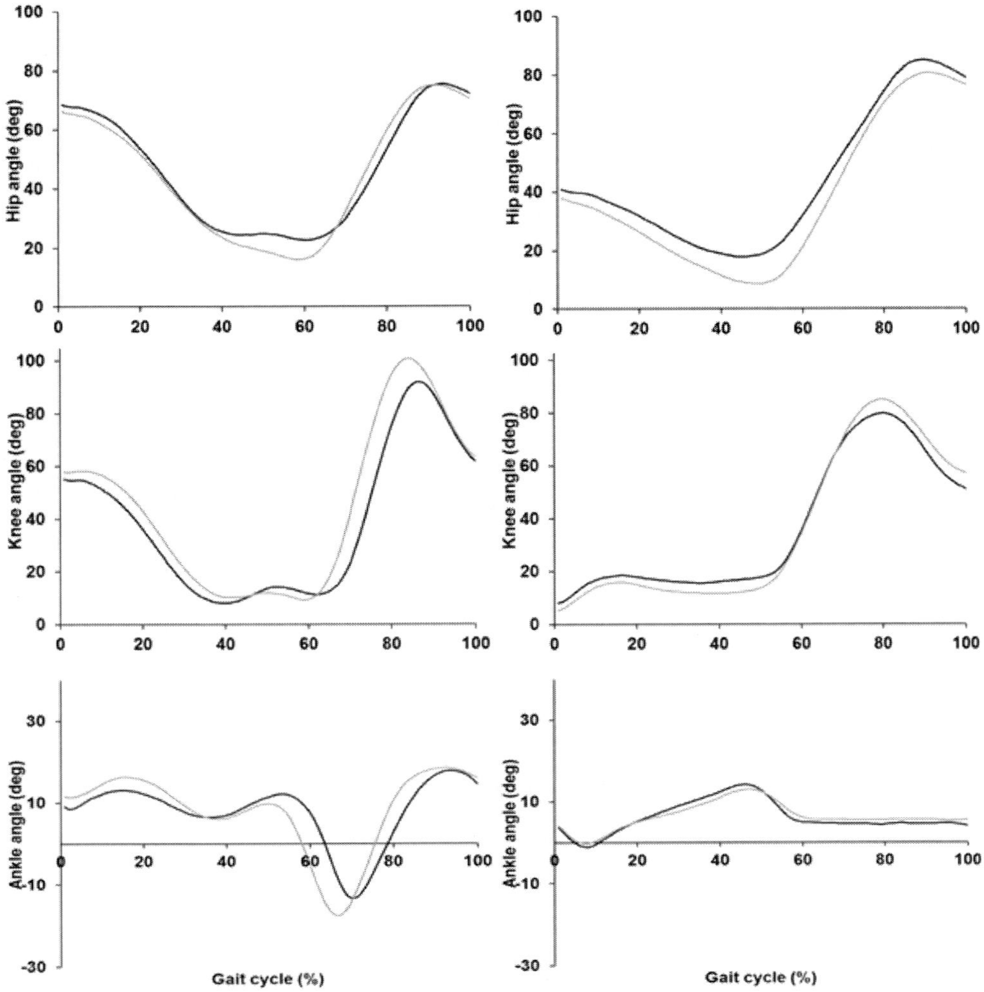

Figure 4. Sagittal plane joint angles of the A) lead (unaffected) and B) trail (amputated) limbs during stair ascent. Averaged data are presented. Amputee non-fallers (black line); amputee fallers (grey line). The gait cycle is initiated and terminated with foot contact. Participants exhibited a constant unaffected lead limb preference when ascending the stairs (Vanicek, 2009). Hip flexion, knee flexion and ankle dorsiflexion are positive.

The most functionally demanding phase during stair descent, the controlled lowering phase, characterised by a large knee extensor eccentric power burst during single support and also labelled the K3 power burst (McFadyen and Winter, 1988), is compensated for by a hip strategy with a large hip extensor moment and hip power absorption. Only one published study has investigated gait differences between amputee fallers and non-fallers during stair ascent specifically. More biomechanically challenging tasks than level gait are likely to elicit more obvious differences between these two groups, and these results could have important implications for rehabilitation programmes. Vanicek et al., (2010) reported that amputee fallers negotiated stairs more quickly, which could predispose them towards experiencing a fall especially if they lack the joint ROM and muscle strength necessary to recover from a trip or stumble.

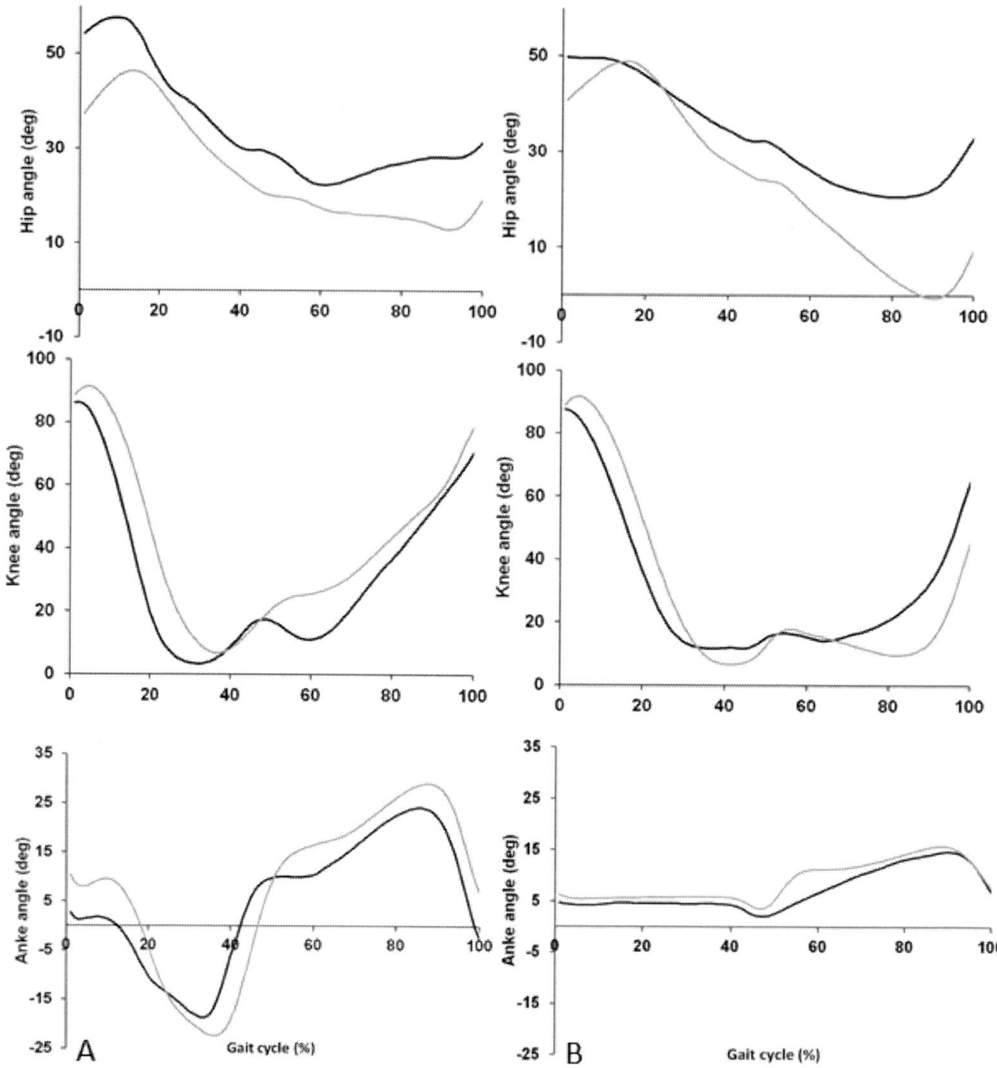

Figure 5. Sagittal plane joint angles of the A) unaffected and B) amputated limbs during stair descent. Averaged data are presented. Amputee non-fallers (black line); amputee fallers (grey line). The gait cycle is initiated and terminated with toe off. On average, swing represents approximately the first 40% of the gait cycle for both groups. Participants did not demonstrate a definite lead limb preference when descending the stairs (Vanicek, 2009). Hip flexion, knee flexion and ankle dorsiflexion are positive.

Amputee non-fallers showed more hip flexion with the unaffected leg and more anterior pelvic tilt during the forward continuance phase (i.e., pre-swing) suggesting they climbed stairs with more forward trunk lean, perhaps as a result of using the handrails more regularly or looking downwards at the feet. This may reveal an important strategy, especially among amputees who lack feedback from the prosthetic foot and leg and thereby use visual information in the absence of proprioceptive and somatosensory input. Interestingly, two participants (both previous fallers) self-selected a step-to gait pattern rather than an alternate, step-over-step pattern, thus walking 35-50% more slowly than the non-fallers and fallers, respectively. Moreover, they were heavily reliant upon the handrails for support. Previous studies have reported that light handrail use improves postural stability (Reeves et al., 2008b) by providing an extra contact point.

To date, no published studies have examined the gait patterns of amputees during stair descent according to falls history specifically. Data from Vanicek (2009) showed that amputee non-fallers walked 31% more slowly than the fallers. This emphasised that walking down stairs at fast speeds could put individuals at risk of falling if they did not have sufficient muscular strength (i.e., eccentric control of the knee extensors during the single support, controlled lowering phase of stair descent) to prevent a fall in case of a trip or stumble. In the same group of transtibial amputees, two participants self-selected a step-to rather than an alternate step-over-step during stair ascent.

However, four participants (2 fallers and 2 non-fallers) chose to descend the stairs using a step-to pattern, most likely as a result of functional and strength limitations at the knee of both limbs. The controlled lowering phase, the most vulnerable instant during stair descent, was substantially shorter for the unaffected (trail) limb and virtually absent for the affected (lead) limb, as the knee on the affected limb did not need to flex eccentrically during single support.

The amputee non-fallers also showed a tendency to 'throw' their prosthetic foot down onto the next step compared to the fallers. This was evidenced by greater hip flexion at toe off and throughout swing, thus lifting the whole leg into the air for stair clearance.

Similar observations were reported in transtibial amputees when crossing obstacles with their prosthesis as the lead limb (Hill et al., 1997). Vanicek (2009) noted the non-fallers appeared to 'throw' their prosthesis over the edge of the step, whilst the fallers employed the more difficult 'roll over' technique. Rolling over the edge of the step is a mechanically challenging task. It requires proper positioning of the prosthetic foot onto the nose of the step and rolling over, whilst in single support (on the affected leg). It is possible that the amputee fallers placed themselves at greater risk of falling by using this technique and relying on eccentric control of the knee extensors during the controlled lowering phase. A previous study demonstrated that handrail use could redistribute some of the work onto the arms and partially unload the legs, thereby reducing the demands on the knee extensors (Reeves et al., 2008b). These findings support the recommendations for interventions aimed at increasing knee extensor eccentric strength and knee joint mobility to improve safe stair descent (Vanicek, 2009). Moreover, using a handrail could reduce the potential for falls, and in the weakest individuals, a step-to stair descent pattern may be encouraged.

Postural Control

Falls are likely to result from inadequate postural control in both able-bodied individuals and lower limb amputees. During quiet stance, balance is controlled by maintaining the body's centre of gravity over the base of support, which necessitates the ability to accurately predict and detect perturbations and then execute appropriate responses. Not realising how far the body's centre of gravity has been displaced, or failure to produce the appropriate response to correct this displacement because of muscle weakness or insufficient joint mobility, will likely result in a fall (Judge et al., 1995). For able-bodied individuals, small perturbations can be corrected by relatively small motions at the ankle joint, termed *ankle strategy*. Lower limb amputees are unable to generate an active torque about the prosthetic ankle joint and thus cannot produce an ankle strategy on the amputated side. Such biomechanical deficits may place greater reliance on the *hip* and/or *stepping strategies* to maintain the centre of gravity within the base of support.

Isakov et al., (1992) reported that transtibial amputees showed greater sway than control subjects when they were asked to stand with their eyes open or eyes closed during their rehabilitation. Similarly, Hermodsson et al., (1994) found that traumatic and dysvascular transtibial amputees had significantly more lateral sway during a two-legged standing test than control subjects and that vascular amputees had significantly more sway than traumatic amputees with and without visual input. Increased lateral sway has been reported in older persons as a possible sign of decreasing balance capacity (Baloh et al., 1995).

These authors also found traumatic amputees had decreased sway in the sagittal direction and this was related to the stiffness of the prosthetic ankle/foot complex (Hermodsson et al., 1994). In lower limb amputees, maintaining balance places additional demands on musculature of the unaffected limb (Isakov et al., 1994; Vrieling et al., 2008b). Transtibial amputees load approximately 62-63% of their body weight through the unaffected limb (Vrieling et al., 2008b) although the passive properties of the prosthetic ankle-foot complex can contribute to postural control (Curtze et al., 2012). Amputees also rely on contributions from the unaffected limb, specifically on increased centre of pressure movement and loading under the unaffected foot, to maintain balance during forward and backward induced falls (Curtze et al., 2012; Vrieling et al., 2008b).

While the residual limb on the amputated side cannot fully compensate for the loss of the anatomical foot (and knee, in the case of transfemoral amputees) as a source of somatosensory feedback, previous research has shown that well-trained transtibial amputees were able to adapt to uncomfortable balancing conditions, implying that prosthetic experience was an important factor in postural control (Isakov et al., 1994). Other research has noted that less able and/or experienced amputees could benefit from different prosthetic design that facilitates postural control. Curtze et al., (2012) reported that transitibial amputees could have an advantage by wearing a prosthetic foot with a large radius of curvature to improve standing balance.

Computerised dynamic posturography (CDP) using the NeuroCom EquiTest® is an objective measure of the contributions of visual, vestibular and somatosensory input for postural control. This method detects postural sway by measuring shifts in the centre of gravity as a person moves within their limits of stability and can quantify postural strategies to static and dynamic perturbations. Quantitative posturography has been used previously to assess balance with elderly (Camicioli et al., 1997; Horak et al., 1989; Judge et al., 1995) and differentiate between elderly fallers and non-fallers (Parry et al., 1995; Wallmann, 2001). Despite the popularity of the NeuroCom EquiTest® to measure balance in a clinical setting, there is a paucity of research that has assessed the use of CDP with lower limb amputees and only one study has specifically compared amputee fallers with non-fallers.

Kaufman and colleagues (2007) used the Sensory Organisation Test (SOT) of the NeuroCom EquiTest® to quantify functional postural control of transfemoral amputees fitted with a mechanical (Mauch SNS) or micro-processor controlled (C-Leg, OttoBock) knee joint. Their findings revealed that participants' balance improved significantly across all six conditions of the SOT when fitted with the micro-processor knee. Jayakaran et al., (2011) assessed balance in a group of older (mean age 69.53 years) unilateral transitibal amputees with approximately 12 years of prosthetic experience. Ten of the 15 participants experienced a 'loss of balance' on the first trial in Conditions 4-6. Compared to Conditions 1-3, these are progressively more challenging and dynamic tests, in which the support surface and visual surround are sway-referenced and move according to the individual's anterior-posterior sway. But no participant physically fell to the ground, as they wore a safety harness preventing them from actually falling. These authors reported a shift from use of the ankle strategy in the relatively simple static balance tasks to reliance on the hip strategy during more dynamic balance tasks, particularly in Conditions 5 and 6 which had altered visual and somatosensory input. Their findings were in agreement with those reported by Vanicek et al., (2009b), the only study that has investigated postural differences between fallers and non-fallers in lower limb amputee.

Vanicek et al., (2009b) assessed postural control using the SOT, but also with the Motor Control Test (MCT) on the NeuroCom EquiTest®. These authors reported that amputee fallers actually scored significantly better than the non-fallers on the Equilibrium Score in Condition 6, which indicated that amputee non-fallers might rely too heavily on visual information in dynamic conditions and that this might actually predispose them towards experiencing a future fall. There were no significant differences in Strategy Scores between the amputee fallers and non-fallers, suggesting that use of the ankle or hip strategy was not able to differentiate previous fallers from non-fallers. Findings from the MCT showed that amputee fallers bore significantly more weight through their amputated vs. unaffected limb during backwards and forwards translation movements, even though the fallers displayed better inter-limb symmetry (i.e., their body weight was more equally distributed across both limbs). In lower limb amputees, particularly in those with inadequate muscle strength to rectify rapid shifts in COP, it appears that loading the amputated limb more, perhaps in an attempt to maintain symmetry between the two limbs, may actually have detrimental effects on balance. These findings have implications for rehabilitation programmes, as weight distribution and standing exercises could be performed in a clinical setting with the use of simple bathroom scales or with the more recent popularity of exercise game consoles, such as the Wii Fit.

Game Console Use As Novel Training Tool

Balance training is an important aspect for gait training and rehabilitation in lower limb amputees. Postural control can be measured using sophisticated equipment, such as the force platform or with computerised dynamic posturography. However, this equipment is expensive and involves complicated analyses and thus is not practical for the clinical environment. To overcome this, many therapists and health care practitioners use clinical tests, such as the Berg Balance Scale (Berg et al., 1989) and the Tinetti Mobility Test (Tinetti et al., 1986), to assess balance performance in both static and dynamic conditions. However, these tests are subjective, prone to ceiling effects and unable to detect small but meaningful improvements. Therefore, relatively inexpensive and portable tools that can measure balance objectively would be a valuable asset for the clinical environment.

Recently, researchers have investigated the novel use of the Wii Balance Board, part of the commercially-available Nintendo video game, as a balance training tool. The Wii Balance Board is equipped with 4 strain gauge load sensors to measure centre of gravity and to track movements during weight-shifting. Participants make subtle adjustments to their body positions using interactive games and are given instantaneous feedback on the television screen. Training with the Wii Balance Board can make rehabilitation and exercise more enjoyable for the participant, whilst the therapist benefits by collecting and analysing laboratory-grade balance data (Clark et al., 2010). There are few empirical studies that have evaluated the effects of a Wii Balance Board, but recent research has concluded that the Wii Balance Board showed excellent centre of path length test-retest reliability and concurrent validity when compared against a gold standard force platform. The Wii Balance Board is not a direct replacement for a force platform in activities that emphasise rapid and high force movements, but it provides comparable data to a force platform when assessing COP path length during standing balance trials.

Several pilot studies have demonstrated promising preliminary results with the use of the Wii Balance Board for improving gait and balance in older adults. Agmon et al., (2011) reported that a 3-month, limited supervision, home-based balance training programme using the Wii Balance Board was safe, realistic and enjoyable for older adults with impaired balance.

Moreover, participants significantly improved their Berg Balance Score and increased their gait speed following the intervention. More recently, Padala et al., (2012) demonstrated the Wii-Fit programme significantly improved Berg Balance Score and Tinetti Mobility Test score in a group of participants in an assisted living environment with mild Alzheimer's Dementia. These participants completed supervised strengthening, yoga and balance exercises using the Wii-Fit 5 times a week for 30 minutes over 8 weeks.

Such favourable results using a novel training tool have also sparked the interest of physiotherapists working in amputee rehabilitation, including some of the prosthetics' manufacturers. In the UK, the British Association of Chartered Physiotherapists in Amputee Rehabilitation (BACPAR), a working group of the Chartered Society of Physiotherapists, released guidelines for the safe use of the Wii Fit among lower limb prosthetics wearers (BACPAR, 2009).

Together with the prosthetics' manufacturer, Őssur, movements performed on the Wii Fit provided an enjoyable environment in which to practice postural control and enhance proprioceptive feedback and sensory awareness. Specifically, the balance movements were effective at enhancing postural alignment, weight transfer and body control; the muscle workout exercises targeted core stability; the yoga movements incorporated important lower limb stretches for amputee limb wearers; and the aerobic exercises facilitated a fun way to increase cardiovascular fitness (Össur, 2009).

No empirical data have been published specifically for lower limb amputees training with the Wii Fit; however, research is currently underway to investigate the effects of Wii Fit console use on balance performance and falls in this population. Anecdotal feedback suggests incorporating this mode of training is beneficial by motivating the patient and providing visual biofeedback that might not otherwise be possible in a clinical setting.

Conclusion

Lower limb amputees are at a particularly high risk of experiencing a fall, especially new amputees. Appropriate and targeted exercise interventions, as part of the prosthetic rehabilitation phase and beyond, are needed to address reduced joint mobility, muscle strength deficits and muscle tightness, mainly during more demanding activities of daily. The findings from many studies to date advocate exercises specifically aimed at improving knee extensor eccentric strength bilaterally in the case of transitibial amputees, and hip musculature in both transtibial and transfemoral amputees. Moreover, some sub-phases of the gait cycle, such as the transition from double support into single support on the affected side, have been highlighted as particularly vulnerable instances during many modes of walking. Performing movements too quickly, especially amongst relatively recent amputees or those without adequate strength or joint mobility, may be considered more risky and caution should be taken. Where possible (on stairs), handrail use should be encouraged.

Interpreting somatosensory feedback from the residual limb and remaining joint(s) and musculature could provide important sensory information in the absence of the biological knee and/or ankle-foot. Over-reliance on visual information is not advocated, as it can sometimes be inaccurate. These recommendations should be incorporated into standard physiotherapy practice and home-based exercise programmes.

References

Agmon, M., Perry, C. K., Phelan, E., Demiris, G., Nguyen, H. Q. (2011). A pilot study of Wii Fit exergames to improve balance in older adults. *Journal of Geriatric Physical Therapy,* 34: 161-167

American Geriatrics Society (AGS) (2001). Panel on Falls Prevention. Guideline for the prevention of falls in older persons. *Journal of the American Geriatrics Society,* 49: 664-672

Aldridge, J. M., Sturdy, J. T., Wilken, J. M. (2011). Stair ascent kinematics and kinetics with a powered lower leg system following transtibial amputation. *Gait and Posture,* 36: 291–295

Alimusaj, M., Fradet, L., Braatz, F., Gerner, H. J., Wolf, S. I. (2009). Kinematics and kinetics with an adaptive ankle foot system during stair ambulation of transtibial amputees. *Gait and Posture,* 30: 356–363

Askham, J., Glucksman, E., Owens, P., Swift, C., Tinker, A., Yu, G. (1990). *A review of research on falls among elderly people*. London King's College: Age Concern Institute of Gerontology.

Bae, T. S., Choi, K., Mun, M. (2009). Level walking and stair climbing gait in above-knee amputees. *Journal of Medical Engineering and Technology,* 33: 130–135.

Baloh, R. W., Spain, S., Socotch, T. M., Jacobson, K. M., Bell, T. (1995). Posturography and balance problems in older people. *Journal of the American Geriatrics Society,* 43: 638-644

Barnett, C., Vanicek, N., Polman, R., Hancock, A., Brown, B., Smith, L., Chetter, I. (2009). Kinematic gait adaptations in unilateral transtibial amputees during rehabilitation. *Prosthetics and Orthotics International,* 33: 141-53

Beaulieu, F. G. D., Pelland, L., Robertson, D. G. E. (2008). Kinetic analysis of forwards and backwards stair descent. *Gait and Posture,* 27: 564–571

Berg, K., Wood-Dauphinee, S., Williams, J. I., Gayton, D. (1989). Measuring balance in the elderly: Preliminary development of an instrument. *Physiotherapy Canada,* 41: 304-311

Buczek, F. L., Cavanagh, P. R., Kulakowski, B. T., Pradhan, P. (1990). Slip resistance needs of the mobility disabled during level and grade walking. In: Gray, B. E., editor. *Slips, stumbles and falls: pedestrian footwear and surfaces, ASTM STP 1103.* Philadelphia: American Society for Testing and Materials. p. 39–54.

Camicioli, R., Panzer, V. P., Kaye, J. (1997). Balance in the healthy elderly: posturography and clinical assessment. *Archives of Neurology,* 54: 976-981.

Clark, R. A., Bryant, A. L., Pua, Y., McCrory, P., Bennell, K., Hunt, M. (2011). Validity and reliability of the Nintendo Wii Balance Board for assessment of standing balance. *Gait and Posture.* 31: 307–310

Curtze, C., Hof, A. L., Postema, K., Otten, B. (2012). The relative contributions of the prosthetic and sound limb to balance control in unilateral transtibial amputees. *Gait and Posture.* 36: 276-281

DeVita, P., Hortobagyi, T. (2000). Age causes a redistribution of joint torques and powers during gait. *Journal of Applied Physiology* 88: 1804-1811

Esquenazi, A. (2004). Amputation rehabilitation and prosthetic restoration. From surgery to community reintegration. *Disability and Rehabilitation* 26: 831-836

Glaister, B. C., Bernatz, G. C., Klute, G. K., Orendurff, M. S. (2007). Video task analysis of turning during activities of daily living. *Gait and Posture.*25: 289–294

Hermodsson, Y., Ekdahl, C., Persson, B. M., Roxendal, G. (1994). Standing balance in transtibial following vascular disease or trauma: A comparative study with healthy subjects. *Prosthetics and Orthotics International.* 18: 150-158

Hill, S. W., Patla, A. E., Ishac, M. G., Adkin, A. L., Supan, T. J., Barth, D. G. (1997). Kinematic patterns of participants with a below-knee prosthesis stepping over obstacles of various heights during locomotion. *Gait and Posture.* 6: 186-192

Hobara, H., Kobayashi, Y., Nakamura, T., Yamasaki, N., Nakazawa, K., Akai, M., Ogata, T. (2011). Lower extremity joint kinematics of stair ascent in transfemoral amputees. *Prosthetics and Orthotics International* 35: 467–472

Horak, F. B., Shupert, C., Mirka, A. (1989). Components of postural dyscontrol in the elderly: a review. *Neurobiology of Aging* 10: 727-738.

Information Services Division, NHS Scotland: National Amputee Statistical Database (NASDAB). *The Amputee Statistical Database for the UK – Annual Report 2006/07.* Edinburgh: National Amputee Statistical Database; 2009

Isakov, E., Mizrahi, J., Ring, H., Susak, Z., Hakim, N. (1992). Standing sway and weight-bearing distribution in people with below-knee amputation. *Archives of Physical Medicine and Rehabilitation* 73: 174-178

Isakov, E., Mizrahi, J., Susak, Z., Ona, I., Hakim, N. (1994). Influence of prosthesis alignment on the standing balance of below-knee amputees. *Clinical Biomechanics* 9: 258-262

Jayakaran, P., Johnson, G. M., Sullivan, S. J. (2011). Test-retest reliability of the Sensory Organization Test in older persons with a transtibial amputation. *PM and R* 3: 723-729.

Johansson, J. L., Sherrill, D. M., Riley, P. O., Bonato, P., Herr, H. (2005). A clinical comparison of variable-damping and mechanically passive prosthetic knee devices. *American Journal of Physical Medicine and Rehabilitation* 84: 563-575

Jones, S. F., Twigg, P. C., Scally, A. J., Buckley, J. G. (2006). The mechanics of landing when stepping down in unilateral lower-limb amputees. *Clinical Biomechanics* 21: 184–193

Judge, J. O., King, M. B., Whipple, R., Clive, J., Wolfson, L. I. (1995). Dynamic balance in older persons: Effects of reduced visual and proprioceptive input. *The Journals of Gerontology. Series A, Biological Science and Medical Sciences* 50A: M263-270.

Kaufman, K. R., Levine, J. A., Brey, R. H., Iverson, B. K., McCrady, S. K., Padgett, D. J., Joyner, M. J. (2007). Gait and balance of transfemoral amputees using passive mechanical and microprocessor-controlled prosthetic knees. *Gait and Posture* 26: 489–493

Kerrigan, D. C., Todd, M. K., Della Croce, U., Lipsitz, L. A., Collins, J. J. (1998). Biomechanical gait alterations independent of speed in the healthy elderly: Evidence for specific limiting impairments. *Archives of Physical Medicine and Rehabilitation* 79: 317-322

Kerrigan, D. C., Lee, L. W., Nieto, T. J., Markman, J. D., Collins, J. J., Riley, P. O. (2000). Kinetic alterations independent of walking speed in elderly fallers. *Archives of Physical Medicine and Rehabilitation* 81: 730-735

Lin, H. C., Lu, T. W., Hsu, H. C. (2005). Comparisons of joint kinetics in the lower extremity between stair ascent and descent. *Journal of Mechanics* 21: 41-50

Mattes, S. J., Martin, P. E., Royer, T. D. (2000). Walking symmetry and energy cost in persons with unilateral transtibial amputations: Matching prosthetic and unaffected limb inertial properties. *Archives of Physical Medicine and Rehabilitation* 81: 561-568

Mazari, F. A. K., Mockford, K. A., Barnett, C., Khan, J. A., Brown, B., Smith, L., Polman, R. C., Hancock, A., Vanicek, N., Chetter, I. C. (2010). Hull early walking aid for rehabilitation of transtibial Amputees - Randomised controlled Trial (HEART). *The Journal of Vascular Surgery* 52: 1564-1571

McFadyen, B. J., Winter, D. A. (1988). An integrated biomechanical analysis of normal stair ascent and descent. *Journal of Biomechanics* 21: 733–744

Miller, W. C., Speechley, M., Deathe, A. B. (2001). The prevalence and risk factors of falling and fear of falling among lower extremity amputees. *Archives of Physical Medicine and Rehabilitation* 82: 1031-1036

National Limb Loss Information Centre (NLLIC). *Amputation Statistics by Cause Limb Loss in the United States*. Knoxville, TN: Amputee Coalition of America. 2008. Retrieved on 23 August 2012 from http://www.amputee-coalition.org/fact_sheets/amp_stats_cause.html

Össur. Wii Fit for lower limb prosthetic users. Manchester, UK: Össur UK. 2009 Retrieved on 30 August 2012 from http://www.wiihabilitation.co.uk/files/wii_fit_bro_uk_traffic_light_version.pdf

Padala, K. P., Padala, P. R., Malloy, T. R., Geske, J. A., Dubbert, P. M., Dennis, R. A., Garner, K. K., Bopp, M. M., Burke, W. J., Sullivan, D. H. (2011). Wii-Fit for improving gait and balance in an assisted living facility: A pilot study. *Journal of Aging Research* Article ID 597573

Parry, A., Anderson, A., Zinober, B. (1995). *Evaluation of dynamic posturography with the EquiTest system: a two-month pilot study to determine the feasibility of assessment of sensory organization and motor control to predict propensity to fall in elderly people*. Sheffield: Sheffield Hallam University, Health Research Institute.

Parry, M., Morrison, J. D. (1989). Use of the Femurett adjustable prosthesis in the assessment and walking training of new above-knee amputees. *Prosthetics and Orthotics International* 13: 36-38

Powers, C. M., Boyd, L. A. (1997). Stair ambulation in persons with transtibial amputation: An analysis of the Seattle LightFoot. *Journal of Rehabilitation Research and Development* 34: 9-18

Reeves, N. D., Spanjaard, M., Mohagheghi, A. A., Baltzopoulos, V., Maganaris, C. N. (2008a) The demands of stair descent relative to maximum capacities in elderly and young adults. *Journal of Electromyography and Kinesiology* 18: 218–227

Reeves, N. D., Spanjaard, M., Mohagheghi, A. A., Baltzopoulos, V., Maganaris, C. N. (2008b). Influence of light handrail use on the biomechanics of stair negotiation in old age. *Gait and Posture* 28: 327–336

Schmalz, T., Blumentritt, S., Marx, B. (2007). Biomechanical analysis of stair ambulation in lower limb amputees. *Gait and Posture* 25: 267–278

Scott, H., Condie, M. E., Treweek, S. P., Sockalingam, S. (2000). An evaluation of the Amputee Mobility Aid (AMA) early walking aid. *Prosthetics and Orthotics International* 24: 39-46

Segal, A. D., Orendurff, M. S., Czerniecki, J. M., Schoen, J., Klute, G. K. (2011). Comparison of transtibial amputee and non-amputee biomechanics during a common turning task. *Gait and Posture* 33: 41-47

Strike, S. C., Taylor, M. J. D. (2009). The temporal–spatial and ground reaction impulses of turning gait: Is turning symmetrical? *Gait and Posture* 29: 597-602

Talbot, L. A., Musiol, R. J., Witham, E. K., Metter, E. J. (2005). Falls in young, middle-aged and older community dwelling adults: Perceived cause, environmental factors and injury. *BMC Pub. Health* 5: 86.

Taylor, M. J. D., Strike, S. C., Dabnichki, P. (2006). Turning bias and lateral dominance in a sample of able-bodied and amputee participants. *Laterality* 12: 50-63

Tinetti, M. E., Williams, T. F., Mayewski, R. (1986). Fall Risk Index for elderly patients based on number of chronic disabilities. *American Journal of Medicine* 80: 429-434

Vanicek, N., Strike, S., McNaughton, L., Polman, R. (2009a). Gait patterns in transtibial amputee fallers vs. non-fallers: biomechanical differences during level walking. *Gait and Posture* 29: 415-20

Vanicek, N., Strike, S., McNaughton, L., Polman, R. (2009b). Postural responses to dynamic perturbations in amputee fallers vs. non-fallers: a comparative study with able-bodied subjects. *Archives of Physical Medicine and Rehabilitation* 90: 1018-1025

Vanicek, N. *Biomechanical and psychological factors that distinguish fallers from non-fallers: A comparative study of transtibial amputees and able bodied individuals* [PhD thesis]. Hull, UK: University of Hull; 2009 [cited 2012 August 30]. Available from: British Library Electronic Theses Online Service

Vanicek, N., Strike, S., McNaughton, L., Polman, R. (2010). Gait patterns in transtibial amputee fallers vs. non-fallers: biomechanical differences during stair ascent. *Prosthetics and Orthotics International* 34: 399-410

Vickers, D. R., Palk, C., McIntosh, A. S., Beatty, K. T. (2008). Elderly unilateral transtibial amputee gait on an inclined walkway: A biomechanical analysis. *Gait and Posture* 27: 518-529

Vrieling, A. H., van Keeken, H. G., Schoppen, T., Otten, E., Halbertsma, J. P. K., Hof, A. L., Postema, K. (2008a). Uphill and downhill walking in unilateral lower limb amputees. *Gait and Posture* 28: 235-242

Vrieling, A. H., van Keeken, H. G., Schoppen, T., Otten, E., Hof, A. L., Halbertsma, J. P. K., Postema, K. (2008b). Balance control on a moving platform in unilateral lower limb amputees. *Gait and Posture* 28: 222-228

Wallmann, H. W. (2001). Comparison of elderly nonfallers and fallers on performance measures of functional reach, sensory organization, and limits of stability. *The Journals of Gerontology. Series A, Biological Science and Medical Sciences* 56: 580-583.

Winter, D. A., Sienko, S. E. (1988). Biomechanics of below-knee amputee gait. *Journal of Biomechanics* 21: 361-367

Yack, H. J., Nielsen, D. H., Shurr, D. G. (1999). Kinetic patterns during stair ascent in patients with transtibial amputations using three different prostheses. *Journal of Prosthetics and Orthotics* 1999; 11: 57–62

Ziegler-Graham, K., MacKenzie, E. J., Ephraim, P. L., Travison, T. G., Brookmeyer, R. (2008). Estimating the prevalence of limb loss in the United States: 2005 to 2050. *Archives of Physical Medicine and Rehabilitation* 89: 422-429

In: Gait Biometrics
Editors: Li Li and Matthew Holmes
ISBN: 978-1-62948-381-8

Chapter VIII

Concussion and Gait

Thomas Buckley*
Department of Health and Kinesiology, Georgia Southern University,
Statesboro, GA, US

Abstract

There are an estimated 1.6 – 3.8 million concussions which occur annually in the United States. Current assessment batteries, provided baseline healthy data is available, are up to 96% sensitive in the acute identification of a concussion; however, accurately identifying recovery remains elusive. While computerized neuropsychological testing has received the majority of investigations and media attention, assessment of postural control following a concussion remains a key component of a multi-faceted assessment protocol. The traditional postural control test is a static balance error test; however, recent studies have begun to investigation the effectiveness of clinical gait analysis in identifying lingering impairments in postural control following sports-related concussions. The majority of the limited number of studies have been conducted at one research laboratory and been delimited to grade 2 American Academy of Neurology concussions and tested on limited, but specific, dates. These studies have used three primary investigatory models: single task gait, dual-task gait with working memory cognitive challenges, and obstacle crossing/avoidance paradigms. In the acute recovery phase, typically day 2 post-injury, a conservative gait strategy has been identified which includes noted reductions in gait velocity, wider base of support, reduced stride length, and increased step time and during dual task challenges additional restrictions have been noted including; reductions in stride length, anterior velocity of the COM, and COM displacement in the frontal plane. During obstacle avoidance tasks, post-concussion patients increased their toe clearance likely in an effort to increase the safety margin of the task. As time progressed, lingering and persistent impairments in gait continued with the more challenging tasks demonstrating the slowest recovery patterns. By 28 days post-injury, frontal plane kinematic differences were still present suggesting incomplete recovery from the concussion. Future gait related studies need to incorporate within-

* Email: tbuckley@georgiasouthern.edu.

subject baseline data and comparison to typical clinical post-concussion assessment tools to establish gait analysis as a valid and reliable tool.

Concussion Overview

The National Institute of Health has deemed concussions, a mild traumatic brain injury, a major public health disorder (NINDS, 2012). There are an estimated 1.6 – 3.8 million concussions which occur annually in the United States; however, this number likely underestimates the true epidemiological rate as over half of all concussions may go unreported (Jean A. Langlois, Rutland-Brown, & Wald, 2006; M. McCrea, Hammeke, Olsen, Leo, & Guskiewicz, 2004). There is no universally accepted definition of a concussion; however, the 3rd International Consensus Statement on Concussion in Sport (3rd CIS) has defined concussion as, "a complex pathophysiological process affecting the brain, induced by traumatic biomechanical forces" (Paul McCrory et al., 2009). While headache, dizziness, and confusion are the most common symptoms of concussion; concussions can present with a diverse array of symptoms which are generally categorized into somatic, cognitive, and emotional symptoms as well as physical signs such as loss of consciousness (LOC) and post-traumatic amnesia (PTA) (Guskiewicz, Weaver, Padua, & Garrett, 2000; M. McCrea et al., 2003; Paul McCrory, et al., 2009). A common concussion misconception is that LOC and or PTA are required for a concussion to have occurred; however, they typically occur in less than 10% and 25% of concussions respectively (Guilmette & Paglia, 2004; Guskiewicz et al., 2003; Guskiewicz, et al., 2000; M. McCrea, et al., 2003). This wide range of symptoms underlies the unique and individualized nature of each concussion and is the foundation of Stan Herring, co-author of ACSM's concussion consensus statement, oft quoted comment, "If you see one concussion, you've seen one concussion" (Herring et al., 2011).

There are numerous theories which have been designed to explain the neurophysiological process of a concussion (Shaw, 2002). It is generally accepted that a concussion is marked by a neurometabolic cascade initiated by unchecked neuronal depolarization that leads to an efflux of potassium and influx of calcium (Giza & Hovda, 2001). (See Figure 1) This ionic shift produces an energy crisis in the brain due to hyperglycolysis, impaired oxidative metabolism secondary to calcium sequestration in the mitochondria, and lactate accumulation (Giza & Hovda, 2001). This lactate accumulation is one speculated pathophysiological process underlying the risk of second impact syndrome as is increased physiological acidosis, cellular membrane damage, altered blood brain barrier permeability and overall cerebral edema; all of which potentially leave neurons more vulnerable to the second impact (Giza & Hovda, 2001). Based on animal models, this ionic shift lasts 7–10 days, although some evidence suggests that mitochondrial calcium continues to impair oxidative metabolism in the cortex and hippocampus for 10 days or more (Giza & Hovda, 2001; Hovda, Yoshino, Kawamata, Katayama, & Becker, 1991; Shaw, 2002).

Appropriate concussion management is critical to prevent recurrent injury in the short-term and reduce the risk for long term later-life pathologies. Specifically, allowing an acutely concussed individual to prematurely return to participation elevates the risk of the potentially, albeit rare and debated, fatal second impact syndrome (Bey & Ostick, 2009; Cantu, 1998; P. McCrory, 2001; Paul McCrory, Davis, & Makdissi, 2012). While the exact neurophysiological pathology is elusive, second impact syndrome is generally believed to occur when the brain suffers a second impact, while still metabolically recovering from a concussion (Cantu, 1998). This second injury results in a loss of cerebral autoregulation trigging a pathological process which includes rapid cerebral vascular congestion, increased intracranial pressure, brain herniation and often death within minutes (Bey & Ostick, 2009; Cantu, 1998). A review of catastrophic head injuries highlights the risk as almost 60% of football players suffering catastrophic head injuries had a previous head injury and almost 40% admitted to playing despite residual symptoms of the prior head injury (Boden, Tacchetti, Cantu, Knowles, & Mueller, 2007). Subacutely, suffering a single concussion elevates the risk of suffering a future concussion by 3–6 fold with the majority of recurrent injuries occurring within 10 days of the original injury (Guskiewicz, et al., 2003; Michael McCrea et al., 2009; Zemper, 2003). These repeat concussions are likely to present with more symptoms, more likely to experience LOC, and to have delayed recovery (Collins et al., 2002; Iverson, Gaetz, Lovell, & Collins, 2004).

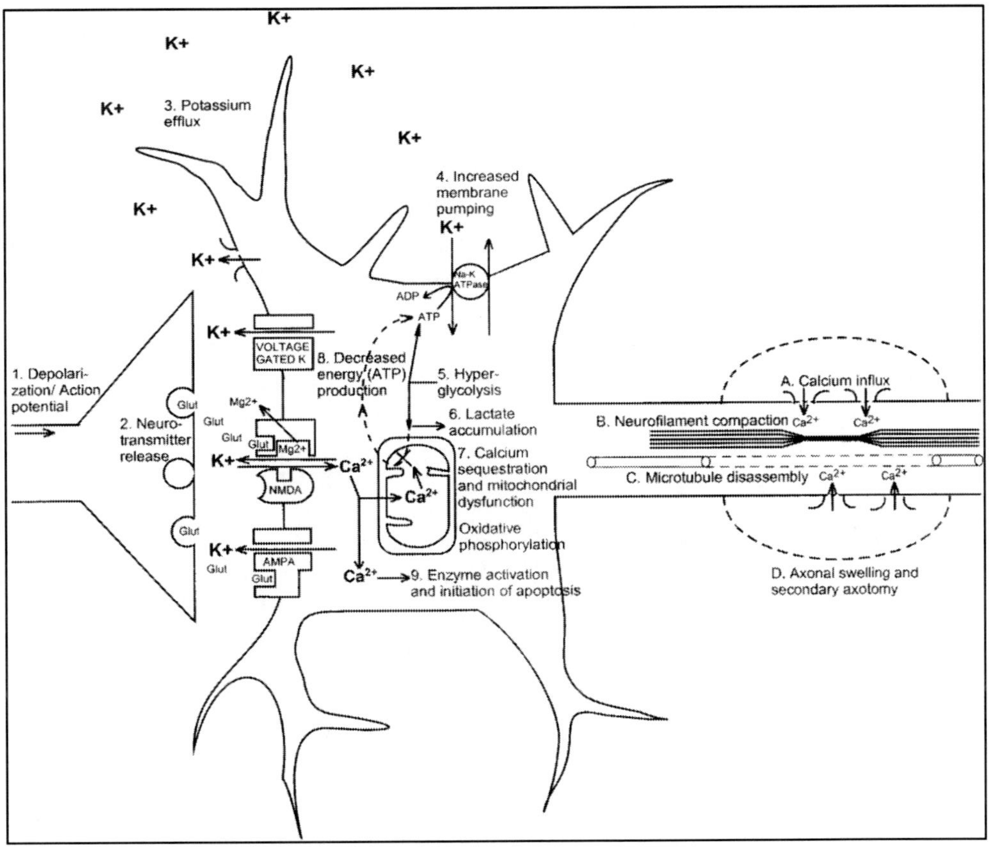

Figure 1. The neurometabolic cascade following a concussion. Giza CC, Hovda DA. The Neurometabolic Cascade of Concussion. *J Athl Train*. 2001 Sep;36(3):228-235.

Recent findings have raised concerns about later-life complications associated with repeated concussions. Retired professional football players who reported a lifetime history of at least 3 concussions had a five-fold increased risk of mild cognitive impairment, a three-fold elevated risk of both significant memory problems and clinically diagnosed depression, and a potential earlier onset of Alzheimer's disease (Guskiewicz et al., 2005; Guskiewicz et al., 2007). It is important to note, however, that these findings were based on retrospective surveys of participants self-reported concussion history, including pre-WWII football players, and a later study reported only moderate reliability of these participants self-reported concussion history (Kerr, Marshall, & Guskiewicz, 2012). Recent, and potentially more disconcerting, findings have linked both concussions and sub-concussive impacts to the later-life development of chronic traumatic encephalopathy in a series of case studies of former football and hockey players, boxers and wrestlers, as well as military veterans exposed to blast injuries (Ann C. McKee et al., 2009; A. C. McKee et al., 2012). Finally, a small series of case studies has suggested a potential link between repeated concussions and the development of amyotrophic lateral sclerosis (Ann C. McKee et al., 2010).

Concussion and Postural Control

Impairments in postural control are a cardinal symptom of a concussion (Paul McCrory, et al., 2009). While debate persists on the structural versus functional nature of a concussion, (P. R. McCrory & Berkovic, 2001; Zhang et al., 2010) impairments in postural control have been speculated to result from neurophysiological changes in the brain's cortex and the brainstem's reticular formation (Guskiewicz, 2011; Shaw, 2002). Based on studies utilizing the sensory organization test, (Figure 2) a sensory interaction deficit, whereby the somatosensory, visual, and vestibular interactions are impaired, has been speculated to be the underlying mechanism of postural control impairments (Guskiewicz, 2011; Guskiewicz, Perrin, & Gansneder, 1996).

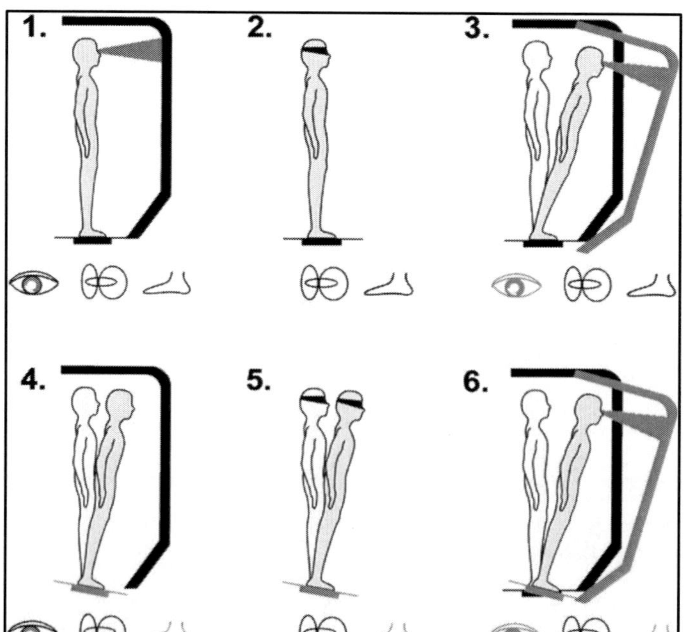

Figure 2. The six stances of the sensory organization test. Guskiewicz KM. Balance Assessment in the Management of Sport–Related Concussion. Clin Sports Med. 2011. 30: 89–102.

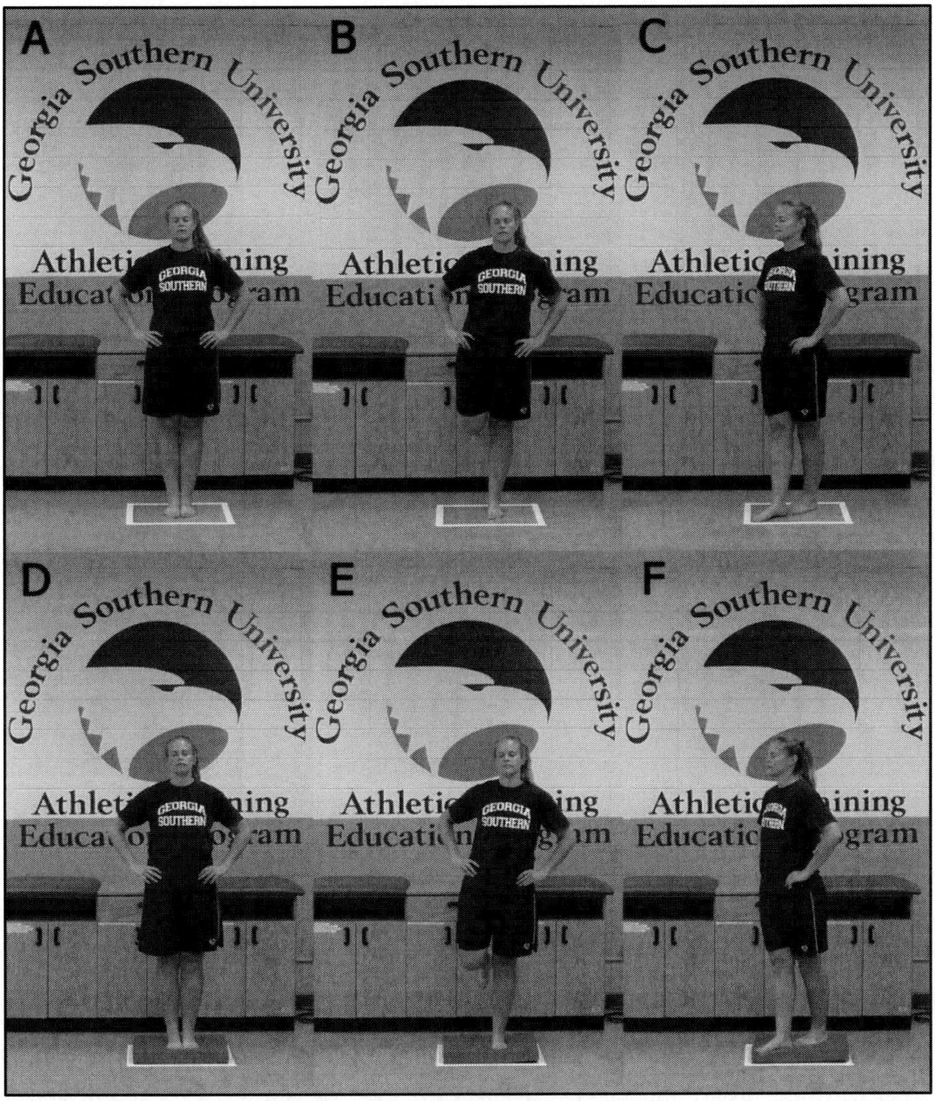

Figure 3. The six stances of the Balance Error Scoring System (BESS) test. Conditions A–C are on a firm surface while conditions D–F are on a foam surface. Conditions A & D have both feet on the surface and in contact, conditions B&E are single leg, and conditions C&F are tandem stance. Each stance is performed for 20 seconds and the total numbers of errors per stance (maximum of 10 per stance) are summed for a total score.

In concussion management, the traditional clinical postural stability assessment tool is the Balance Error Scoring System (BESS) test (Paul McCrory, et al., 2009). (See Figure 3) It consists of performing three stances (double, single, and tandem) on two surfaces (firm and medium density foam) for 20 seconds (Guskiewicz, Ross, & Marshall, 2001). The BESS is scored by recording the number of errors (lifting the hands off of the Iliac Crest, opening their eyes, stepping, stumbling, or falling, moving the hip more than 30° of abduction or flexion, lifting the forefoot or heel, or remaining out of the test position for 5 seconds) committed during each testing component (Riemann, Guskiewicz, & Shields, 1999). Typically, athletes perform it during a preparticipation physical examination to establish a baseline score, and their postconcussion score is compared to the baseline value. Once their postconcussion BESS performance reaches baseline, athletes are presumed to have regained postural stability; which typically occurs within the first 3–5 days post-injury despite the continued presence of concussion related symptoms (Guskiewicz et al., 2004; M. McCrea, et al., 2003; Paul McCrory, et al., 2009). However, multiple studies utilizing more sophisticated technologies have suggested that impairments in postural control can persist for at least 30 days post-injury (Slobounov, Cao, Sebastianelli, Slobounov, & Newell, 2008; Slobounov, Tutwiler, Sebastianelli, & Slobounov, 2006).

Previous investigations suggested good interrater (0.78-0.93) and intrarater (0.87-0.98) reliability (Riemann, et al., 1999; Valovich McLeod et al., 2004). However, a recent report raised concerns over low overall BESS interrater (0.57) and intrarater (0.74) reliability and a minimum detectible change of 9.4 (interrater) and 7.3 (intrarater) errors (Finnoff, Peterson, Hollman, & Smith, 2009). BESS is also limited by previous anaerobic, aerobic, or exertional exercise, testing environment, foot condition (taped, braced, or bare), and, perhaps most important, practice effect (Steven P. Broglio, Sosnoff, Rosengren, & McShane, 2009; Fox, Mihalik, Blackburn, Battaglini, & Guskiewicz, 2008; Onate, Beck, & Van Lunen, 2007; Susco, McLeod, Gansneder, & Shultz, 2004; Valovich, Perrin, & Gansneder, 2003; Wilkins, McLeod, Perrin, & Gansneder, 2004). Furthermore, the BESS protocol measures static postural stability only without consideration for dynamic postural stability during transitional or continuous movements. While potentially useful as a sideline clinical assessment to identify the presence of a concussion, these limitations leave BESS scientifically inadequate to assess to recovery and readiness to return to participation.

Concussion and Gait

Compared to other commonly investigated neurological pathologies (e.g., Parkinson's disease, elderly fallers, stroke, amputee), investigations of gait to identify impairments in postural control post-concussion have been fairly limited. Indeed, there are more review article (e.g., systematic reviews, meta-analyses) on gait in the elderly than original research articles related to concussion and gait. The majority of gait studies were performed at one laboratory and were largely delimited to grade II concussions, as defined by the American Academy of Neurology (no LOC and symptoms persisting longer than 15 minutes), (Kelly & Rosenberg, 1997) had fairly homogeneous and small (n=10 – 17/group) participant populations for most studies, lacked within-subject pre-injury data, and have involved a variety of gait tasks including single task gait, dual task gait with working memory challenges, and obstacle avoidance tasks (R. D. Catena, P. van Donkelaar, & L.-S. Chou, 2007; Robert D. Catena, van Donkelaar, & Chou, 2009, 2011; R. D. Catena, P. van Donkelaar, & L. S. Chou, 2007; R. D. Catena, van Donkelaar, Halterman, & Chou, 2009; Parker, Osternig, & Chou, 2006; Parker, Osternig, Lee, van Donkelaar, & Chou, 2005; Parker, Osternig, van Donkelaar, & Chou, 2007; Parker, Osternig, Van Donkelaar, & Chou, 2006, 2008). Finally, not all raw data is provided for all dependent variables of interest limiting the ability to perform a meta-analysis of the findings.

Early evidence of gait impairments following concussion were reported by McCrory and Berkovic based on video analysis of concussions sustained in the mid 1990's by competitors in Australian rules football (P. R. McCrory & Berkovic, 2000). Gait impairments, operationally defined as ataxic, stumbling, or unsteady gait, was noted post-injury in 41% of concussed athletes with the majority manifesting symptoms immediately post-injury; however, a small percentage, 14%, had a minimal delay of 10 – 20 seconds prior to the onset of gait unsteadiness (P. R. McCrory & Berkovic, 2000). While not specifically studied, it was speculated that gait unsteadiness involved brainstem pathology and was multifactorial including postural tone, cerebellar and labyrinthine function (P. R. McCrory & Berkovic, 2000). This study, while limited to gross video observations without true biomechanical assessment, provided foundational evidence of post-concussion gait impairments.

In the first few days following a concussion, when the BESS score is likely still impaired compared to baseline, post-concussion participants routinely demonstrate a conservative gait strategy which manifests itself as a reduced gait velocity, reduced horizontal separation of the center of pressure (COP) of the stance foot and the whole body center of mass (COM), reduced anterior velocity of the COM, reduced medial-lateral range of motion of the COM, all which reflect a diminished ability to maintain stability in the frontal plane post-concussion. Specifically, across studies there is a mean decrease in gait velocity of 0.128 m/s (10.4%) with the post-concussion subjects, tested within 48 hours of their concussion, having a mean gait velocity of 1.229 m/s (Robert D. Catena, et al., 2007; R. D. Catena, et al., 2007; Parker, et al., 2005; Parker, et al., 2008). Interesting, the often cited minimum healthy gait velocity is typically reported as 1.20 m/s, the velocity typically used to establish crossing times at traffic intersections, (Hoxie, Rubenstein, Hoenig, & Gallagher, 1994; J. A. Langlois et al., 1997; Purser et al., 2005) and the post-concussion subjects only exceeded this value by 0.229 m/s and was well below the normal healthy gait velocity of 1.4–1.6 m/s (Perry, 1992). Further, post-concussion single task stepping characteristics alterations, suggestive of a conservative strategy, include a 7.9% wider base of support, a 7.4cm reduction in stride length, and a 0.05 second increase in step time.

Further, extensive investigations of COM characteristics during single task gait supported the conservative gait strategy hypothesis. Specifically, post-concussion subjects in the sagittal plane have noted reductions in COM displacement, a nearly 1.0m/s reduction in anterior velocity of the COM, and a 2 cm reduction in the anterior displacement of the COM and COP during a stride. In the frontal plane, post-concussion participants increase the medial-lateral displacement of their COM, increase the velocity of the COM in the lateral direction, but do not generate a greater separation of the COP and COM (Robert D. Catena, et al., 2007; R. D. Catena, et al., 2007; Parker, et al., 2005; Parker, et al., 2008). From this series of investigations, the results support the presence of a conservative gait strategy when tested within 48 hours post-concussion.

Utilizing traditional clinical measures of balance (e.g., BESS), large cohort investigators have suggested that postural control returns to its baseline value within 3–5 days post-injury (M. McCrea, et al., 2003). The post-concussion gait studies which have been currently published are limited by lack of within subjects' baseline data; however, they are generally tightly matched to otherwise healthy control subjects. Within this context, gait velocity generally appears to return to a normal value by day 5 or 6 post-injury despite still experiencing concussion related symptoms; (Robert D. Catena, et al., 2009, 2011; Parker, et al., 2007; T. M. Parker, et al., 2006) although in one study it had not recovered by day 28 (Parker, et al., 2008).

This finding, and other similar findings, needs to be taken in context as an apparent practice effect was potentially a confounding variable as the gait velocity steadily increased with each testing session in the healthy control group; however, by day 28 the concussion subjects had still not reached the initial and lowest gait velocity of the control subjects (Parker, et al., 2008). Similar findings were noted in the stepping characteristics (stride time, width, and length) (T. M. Parker, et al., 2006) and sagittal plane COM measurements (anterior displacement and velocity of the COM and the anterior COP-COM separation) (Parker, et al., 2007; T. M. Parker, et al., 2006). However, it is worth noting that in some dependent variables the apparent recovery by day 5 was confounded by a subsequently reduced performance on day 14 (Robert D. Catena, et al., 2009, 2011; T. M. Parker, et al., 2006; Parker, et al., 2008). For example, the anterior separation of the COP and COM was approximately .6cm different between groups at day 5 post concussion; however, the difference more than tripled to almost 2 cm on day 14 (Parker, et al., 2008). Separately, another study found no between group differences on day 5 post-injury in the same variable, but significant differences were noted on both day 14 and day 28 (T. M. Parker, et al., 2006). These findings suggest either potential differences on day 5 are not statistically significant due to small groups and the possibility of the study being underpowered or some residual consequence of delayed impairment following a concussion.

While stepping characteristics and sagittal plane COM kinematics tend to resolve within 5 to 6 days post-injury, the frontal plane kinematics consistently demonstrate delayed recovery and are suggestive of a lingering conservative gait strategy. Frontal plane balance may be a more challenging task for post-concussion individuals as there is a limited base of support during the single support phase of gait (van Donkelaar, Osternig, & Chou, 2006). Further, the balance relationship of the trunk and the swing leg during gait is largely dependent on the active hip abduction moment of the supporting limb (van Donkelaar, et al., 2006). Finally, the horizontal displacement of the stance foot COP and total body COM has been suggested to be a critical factor in determining the external moments at the joints of the supporting limb (van Donkelaar, et al., 2006). During single task gait, post-concussion participants demonstrated limited increases in the medial to lateral COM range of motion and velocity (Robert D. Catena, et al., 2007; R. D. Catena, et al., 2009; Parker, et al., 2005; T. M. Parker, et al., 2006). These impairments appear to persist for up to 28 days post-injury, although similar to the sagittal plane kinematics, intermittent recovery is present (e.g., frontal plane COM displacement is impaired at day 2, 5, and 28 post-injury, but no between groups differences were noted on day 14) (R. D. Catena, et al., 2009; T. M. Parker, et al., 2006).

Taken together, this series of studies suggest that single task gait analysis may be a beneficial addition to identifying impairments in postural control following a sports-related concussion. Specifically, in the immediate aftermath of a concussion, a diverse range of impairments in stepping characteristics, as well as frontal and sagittal plane kinematics are present which suggest a conservative gait strategy has been adopted. Over the course of the first month following the concussion, a gradual recovery has been identified with a general pattern of stepping characteristics being the first to recovery, followed by sagittal plane, and then frontal plane kinematics. However, even 28 days post injury; subtle lingering impairments persist in gait performance with the frontal plane appearing to be most sensitive to ongoing impairments.

Concussion and Dual Task Gait

The dual task paradigm requires an individual to perform a primary task, often a motor task, while simultaneously addressing a cognitive challenge (Lee, Sullivan, & Schneiders, 2012). The capacity to perform a second task during locomotion is highly advantageous as it allows for communication between people, transportation of objects, and manipulation of the environment.

However, performing two tasks simultaneously often degrades performance of one or both tasks and this decrement is present even when the competing tasks do not overlap in either sensory input or motor output modality, suggesting a central source of interference (Hausdorff, Schweiger, Herman, Yogev-Seligmann, & Giladi, 2008; Marois & Ivanoff, 2005; Pashler, 1994; Sharikadze, Cong, Staude, Deubel, & Wolf, 2009). Multiple studies suggest that postural control, spatiotemporal gait parameters, and dynamic stability during locomotion are negatively affected by concomitant performance of both cognitive and motor tasks (Bloem, Valkenburg, Slabbekoorn, & Willemsen, 2001; H. C. Chen et al., 1996; Plummer-D'Amato et al., 2008; Shumway-Cook, Woollacott, Kerns, & Baldwin, 1997; Woollacott & Shumway-Cook, 2002).

A cornerstone of the recent concussion position statements and international consensus statement is the importance of a multifaceted assessment protocol involving both postural control and cognitive challenges (Herring, et al., 2011; P. McCrory et al., 2009). This is of particular importance as no single assessment technique was identified as being more than 70% sensitive; however, a multifaceted approach reached sensitivity levels as high as 96% (S. P. Broglio, Macciocchi, & Ferrara, 2007). In concussion management, the combined assessment of postural control and cognitive processing is likely highly relevant as, individually, they appear to have recovered prior to symptom resolution (M. McCrea, et al., 2003).

This has been suggested to be the result of successful compensation strategies whereby the injured individual is able to successfully recruit additional supraspinal regions; thus allowing them to complete the individual task without obvious external deficits (J. K. Chen et al., 2004). In non-gait concussion studies, it has been suggested that balance control takes precedence over cognitive performance during dual task challenges (S. P. Broglio, Tomporowski, & Ferrara, 2005).

Both impairments in postural control and cognitive processing are known acute consequences of a concussion and thus not surprisingly both are abnormal when tested within 24 hours of the injury. Post-concussion gait studies utilizing a dual-task paradigm have largely focused on utilizing working memory challenges (e.g., reciting the months of the year backwards) from the mini mental examination to assess cognitive performance while performing either level over ground gait or obstacle avoidance (Bell & Hall, 1977). A recent systematic review and meta-analysis by Lee et al, (Lee, et al., 2012) suggested that gait velocity and frontal plane range of motion are sensitive markers of dual task interference in post-concussion individuals. (Figure 4) Specifically, a pooled mean decrease in gait velocity of 0.13 m/s was noted across the meta-analysis which is similar to a noted decrease of 0.17 m/s across a diverse population of neurologically impaired participants (Al-Yahya et al., 2011; Lee, et al., 2012).

a. Concussed

Study or Subgroup	Dual task Mean [m/s]	Dual task SD [m/s]	Dual task Total	Single task Mean [m/s]	Single task SD [m/s]	Single task Total	Weight	Mean Difference IV, Random, 95% CI [m/s]
Catena et al.[15]	1.097	0.166	14	1.219	0.137	14	32.1%	-0.12 [-0.23, -0.01]
Parker et al.[16]	1.097	0.113	10	1.25	0.122	10	38.5%	-0.15 [-0.26, -0.05]
Parker et al.[28] (AT)	1.101	0.174	14	1.227	0.15	14	28.2%	-0.13 [-0.25, -0.01]
Parker et al.[28] (NAT)	1.321	0.114	14	1.27	1.127	14	1.2%	0.05 [-0.54, 0.64]
Total (95% CI)			**52**			**52**	**100.0%**	**-0.13 [-0.20, -0.07]**

Heterogeneity: Tau² = 0.00; Chi² = 0.56, df = 3 (P = 0.90); I² = 0%
Test for overall effect: Z = 4.08 (P < 0.0001)

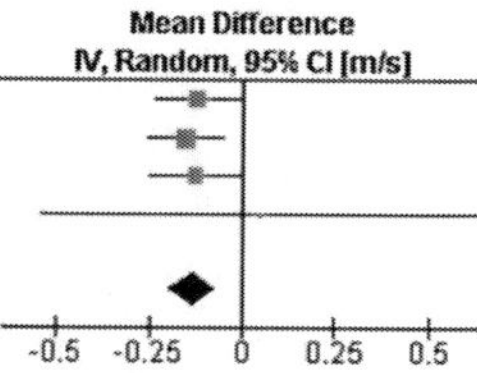

b. Non-concussed

Study or Subgroup	Dual task Mean [m/s]	Dual task SD [m/s]	Dual task Total	Single task Mean [m/s]	Single task SD [m/s]	Single task Total	Weight	Mean Difference IV, Random, 95% CI [m/s]
Catena et al.[15]	1.276	0.133	14	1.361	0.136	14	25.6%	-0.08 [-0.18, 0.01]
Parker et al.[16]	1.219	0.111	10	1.324	0.131	10	22.7%	-0.10 [-0.21, 0.00]
Parker et al.[28] (AT)	1.196	0.152	14	1.217	0.134	14	22.9%	-0.02 [-0.13, 0.09]
Parker et al.[28] (NAT)	1.391	0.142	14	1.381	0.107	14	28.8%	0.01 [-0.08, 0.10]
Total (95% CI)			**52**			**52**	**100.0%**	**-0.05 [-0.10, 0.01]**

Heterogeneity: Tau² = 0.00; Chi² = 3.37, df = 3 (P = 0.34); I² = 11%
Test for overall effect: Z = 1.74 (P = 0.08)

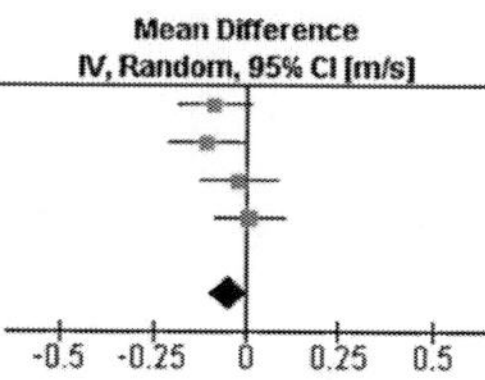

Figure 4. Meta-Analysis of Dual Task. Forest plots for concussed (a) and non-concused (b) groups for gait velocity at day 2. (♦): pooled mean estimate of the differences between ST and DT conditions; (D): difference between ST and DT conditions for individual studies; AT: athletes; NAT: non-athletes; horizontal bars represent 95% confidence intervals. Lee H, Sullivan SJ, Schneiders AG. The use of the dual-task paradigm in detecting gait performance deficits following a sports-related concussion: A systematic review and meta-analysis. J Sci Med Sport. 2012 May 18.

In the acute recovery phase following a concussion, similar to single task gait, multiple impairments are noted during dual task gait. Specifically, when compared to tightly matched control subjects, reductions in stride length, anterior velocity of the COM, and COM displacement in the frontal plane have been reported (Robert D. Catena, et al., 2007, 2009, 2011; R. D. Catena, et al., 2007; Parker, et al., 2005; Parker, et al., 2007; Parker, et al., 2008). Consistent with many dual task paradigms, most kinematic characteristics of gait were reduced with the addition of a cognitive task in both the recently concussed and healthy control groups (Robert D. Catena, et al., 2007, 2009, 2011; R. D. Catena, et al., 2007; Parker, et al., 2005; Parker, et al., 2007; Parker, et al., 2008). The recovery patterns of dual task gait were similar to, but expand upon, the single task gait with apparent lingering deficits still present up to 28 days post-injury (Robert D. Catena, et al., 2009, 2011; Parker, et al., 2007; Parker, et al., 2008). Once again, the frontal plane kinematics appeared most sensitive to the identification of delayed recovery following a concussion (van Donkelaar, et al., 2006). These findings support the necessity of a multifaceted concussion assessment as most cognitive and postural control assessments are recovered far before 28 days post-injury (M. McCrea, et al., 2003). Further, when compared to a commonly utilized computerized neuropsychological testing (Immediate Post-Concussion Assessment and Cognitive Testing or ImPACT), there was little relationship with the dual task gait performance (Parker, et al., 2007). The authors speculated that dynamic motor tasks, such as dual task gait, are potentially more complex and challenging than traditional computerized neuropsychological tests and may better approximate the demands experienced during sports participation (Parker, et al., 2007).

Similar to the single task gait studies, the majority of dual task investigations were performed at one laboratory, lacked within subject baseline data, utilized age and gender matched control subjects, and were delimited to AAN grade 2 concussions (Robert D. Catena, et al., 2007, 2009, 2011; R. D. Catena, et al., 2007; R. D. Catena, et al., 2009; Tonya M. Parker, et al., 2006; Parker, et al., 2005; Parker, et al., 2007; T. M. Parker, et al., 2006; Parker, et al., 2008). A further, and important, methodological note is that the post-concussion studies did not instruct the individuals to prioritize either the postural or cognitive challenge. Notwithstanding these limitations, these findings support the recent and emerging research into dual task challenges in both post-concussion, chronic, and multiple concussions individuals (S. P. Broglio, et al., 2005; Resch, May, Tomporowski, & Ferrara, 2011; Teel, Register-Mihalik, Troy Blackburn, & Guskiewicz, 2012). Current investigations are focused on technologically sophisticated equipment (e.g., sensory organization test, computerized cognitive challenges) which is strongly supported in the laboratory sciences, but lack clinical application. Future studies, in addition to addressing the previously described study limitations, in this area need to transition from the laboratory to the clinic to allow health care providers, without access to sophisticated laboratory technologies, to provide a foundation for the practical application of dual task assessments.

Concussion and Gait Related Challenges

The studies of single and dual task gait have consistently identified a conservative gait strategy being employed by the individuals post-concussion; however, increasing the difficulty of the gait related task may further identify altered postural control strategies. One commonly utilized task in the literature to further challenge the motor component of postural control in elderly individuals and those with neurological diseases is obstacle avoidance or obstacle crossing (Galna, Peters, Murphy, & Morris, 2009; Kovacs, 2005). There are multiple mechanisms which have been suggested to underlie impaired postural control during obstacle crossing including; age-related deficits in vision, proprioception, visual-spatial cognition and attention as well as reductions in internal moments across the hip and ankle during key events in the obstacle crossing gait cycle (Galna, et al., 2009; Kovacs, 2005). These alterations, specifically related to attention, are of particular interest in post-concussion postural control.

An early study on individuals with moderate traumatic brain injury (TBI) suggested that obstacle avoidance may be an important maker of impaired postural control as degraded performance was noted for up to 2 years post-injury (L. S. Chou, Kaufman, Walker-Rabatin, Brey, & Basford, 2004). Specifically, this population demonstrated a significantly reduced displacement and velocity of the COM in the frontal plane and maintained a significantly greater horizontal separation distance of their COP and COM (L. S. Chou, et al., 2004). While increased separation of the COP and COM is often considered an indication of postural control (i.e., reductions in COP – COM separation is often noted during transitional movement tasks in individuals with neurological diseases such as Parkinson's(Hass, Waddell, Fleming, Juncos, & Gregor, 2005; Hass, Waddell, Wolf, Juncos, & Gregor, 2008)), during obstacle avoidance an increase in the horizontal displacement has consistently been shown to associated with impaired postural control (L.-S. Chou, Kaufman, Brey, & Draganich, 2001; L. S. Chou, et al., 2004; Stegemoeller et al., 2012). Further, in the moderate TBI population, the increased displacement of the COM in the frontal plane was independent of the gait velocity (L. S. Chou, et al., 2004; van Donkelaar, et al., 2006). These studies provide justification for the examination of an obstacle avoidance task following a concussion.

Similar to both single and dual task gait studies, the obstacle avoidance investigations have been performed at one laboratory and include the same limitations as previously discussed (Robert D. Catena, et al., 2007, 2009; R. D. Catena, et al., 2009). In the acute recovery stage of a concussion, obstacle avoidance tasks are associated with decreased gait velocity, likely associated with increased stride time, and greater displacement of the COM in the sagittal plane, likely due to the physical constraints of the obstacle (Robert D. Catena, et al., 2007). The concussion individuals increased their toe clearance height over the obstacle, a strategy typically associated with a "safety mechanism", however this may also further challenge the already stressed postural control systems (Robert D. Catena, et al., 2007). Further, obstacle avoidance tasks were associated with greater frontal plane displacement and velocity as well as increased COP and COM separation; however, this occurred for both healthy and concussed individuals (Robert D. Catena, et al., 2007). However, there have been inconsistent findings in these results as other studies have not identified acute changes in obstacle following a concussion (Robert D. Catena, et al., 2009). Interestingly, one possible explanation provided for these findings was that the mental fogginess, a common post-concussion symptom, prevented the individual from recognizing the potential hazard the obstacle presented and thereby they did not alter their crossing strategy (Robert D. Catena, et al., 2009).

As the individual recovers from a concussion, a conservative gait strategy appears to persist while performing an obstacle crossing task for up 28 days post-injury. Specifically, post-concussion individuals reduced the peak velocity of the COM in the frontal plane and reduced horizontal separation of the COP-COM (Robert D. Catena, et al., 2009). It is interesting to note that in the acute recovery phase, there was an increase in COP-COM separation, but then a reduction over time (Robert D. Catena, et al., 2009). This finding supports the presented hypothesis that the obstacle crossing strategy changed over time following a concussion, perhaps because the individual is now more aware of the potential hazard and thus self-selected a more conservative or safer crossing strategy. Finally, one study combined obstacle avoidance with dual task following a concussion (R. D. Catena, et al., 2009). Under these taxing conditions, an altered crossing strategy was noted 28 days post-injury; specifically, there was an increased lead foot clearance and more trial variability, both suggestive of a conservative crossing strategy. Interestingly, on day 2 post-injury, during the acute recovery phase, the post-concussion group had a lower lead limb crossing clearance, although not statistically different, than the healthy control group. While not statistically analyzed in the study, about half of post-concussion participants contacted the obstacle as opposed to only one control subject despite the increased clearance height. Overall, these studies suggest that increasing both the motor and cognitive challenges may further elucidate impairments in postural control following a concussion; although further study is needed to identify why the post-concussion strategy appears to change from the acute stage to a month later.

A single subject investigation of an elite junior hockey player combined obstacle avoidance with a cognitive challenge (Fait, McFadyen, Swaine, & Cantin, 2009). However, unlike most obstacle avoidance studies in which the individual steps over an obstacle set at a specified height, in this study, in an effort to make the motor challenge more sport realistic, the individual had to circumvent a 181 cm tall and 43 cm diameter foam cylindrical obstacle. Despite full resolution of self-reported symptoms and successful completion of the traditional clinical assessment, including computerized neuropsychological testing, the individual continued to demonstrate decreased gait velocity and decrease his clearance around the obstacle as well as continuing to commit errors in the cognitive challenge. While a single subject design has inherent limitations, the results agree with numerous previous studies all suggesting lingering impairments in postural control.

Postural control has also been frequently assessed in the neurological literature through transitional movement paradigms such as gait initiation and termination (Hass et al., 2004; Hass, Waddell, et al., 2008; Hass, Wikstrom, Fournier, Inamdar, & Bishop, 2008; E. A. Wikstrom, Bishop, Inamdar, & Hass, 2010; Erik A. Wikstrom & Hass, 2012). In concussion management, only gait termination has been received preliminary investigation (E. A. Wikstrom, Tapia-Lovler, Munkasy, & Buckley, 2012). The early results suggest that gait termination may be a sensitive tool in identifying impairments in postural control. Specifically, post-concussion gait termination performance was characterized by relative increases in both the propulsive force of the penultimate limb and the braking force of the termination limb when normalized to the gait velocity. Further, these differences persisted for 10 days post-injury despite gait velocity being similar between groups and the post-concussion subjects having fully recovered all traditional measures of clinical assessment. This preliminary finding supports the use of transitional movement paradigms to assess postural control following a concussion.

Finally, while not specifically gait related, post-concussion assessment of postural control has received limited study with static stance using various entropy measures (Cavanaugh et al., 2006; Cavanaugh, Guskiewicz, & Stergiou, 2004, 2005; Gao, Hu, Buckley, White, & Hass, 2011). The initial studies utilized approximate entropy as a measure as it had been established as a clinically useful tool in healthy populations (Cavanaugh, et al., 2005). These studies were limited to 20 second data collections as they were performed in conjunction with the sensory organization test and were only tracked for 96 hours post-injury (Cavanaugh, et al., 2006). This study reported that, specifically in medial-lateral COP oscillations, entropy did not return to baseline values within 96 hours despite apparent recovery on the sensory organization test. Interestingly, the results also indicated that the post-concussion individuals demonstrated reduced entropy, less randomness, than their baseline values and the authors speculated that optimal postural control may be achieved via relatively unconstrained/irregular motor output patterns (Cavanaugh, et al., 2006). More recently, Gao et al, furthered these investigations by applying Shannon and Renyi entrophies to investigate post-concussion postural control and suggested that these methods may have advantageous properties compared to approximate entropy (Gao, et al., 2011). Further, the duration of data collection appears to be an important methodological concern as differences in Shannon entropy were more pronounced after 120 seconds of data; whereas, after only 20 seconds there were no apparent differences. (Figure 5) Finally, the impairments in postural control appeared to persist for at least 10 days post-injury when utilizing the Renyi entropy measure (Gao, et al., 2011). (Figure 6)

Taken together, the results of obstacle avoidance, gait termination, and entropy based studies suggest that continued sophisticated assessment, preferably including within-subject baseline data, of post-concussion postural control is needed to further the scientific understanding of recovery. Additionally, these results further argue that the BESS test, despite being recommended by the 3rd CIS, is wholly insufficient in identifying lingering and persistent impairments in postural control following a concussion.

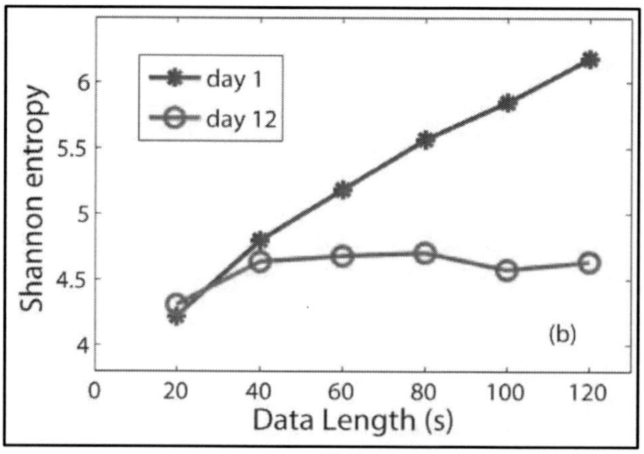

Figure 5. Shannon Entropy measures across time. There were no differences between Day 1 and Day 12 post-injury when the data was sampled for less than 60 seconds. Gao J, Hu J, Buckley TA, White K, Hass CJ. Shannon and Renyi entropies to classify effects of Mild Traumatic Brain Injury on postural sway. PLoS One. 2011;6(9):e24446.

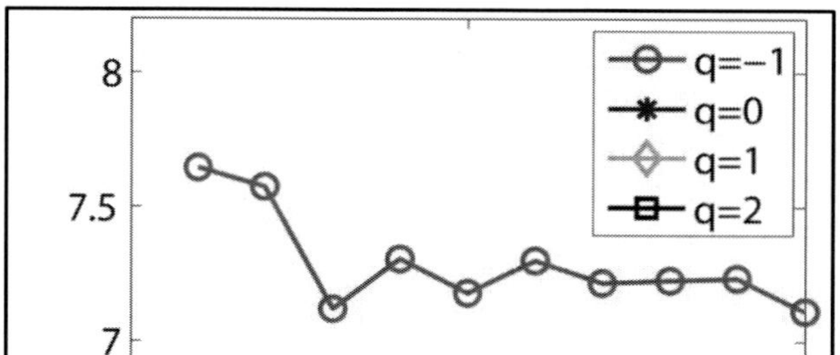

Figure 6. Entropy measures of recovery. Although healthy baseline data is not available for comparison, a clear trend of recovery is present. All participants had achieved baseline values on all clinical assessments prior to day 10. Gao J, Hu J, Buckley TA, White K, Hass CJ. Shannon and Renyi entropies to classify effects of Mild Traumatic Brain Injury on postural sway. PLoS One. 2011;6(9):e24446.

Long–Term Postural Control Changes Post–Concussion

The previous studies, taken in summation, suggest that in the immediate aftermath of a sports-related concussion, impairments in postural control persist for at least up to one month. However, while multiple concussions have been associated with long term neuropathologies, (Daneshvar et al., 2011) there have been limited investigations of gait in this population. This may be a potentially important oversight as the interaction between the normal aging process, (HSU ref) associated with gait impairments and fall risk particularly when a cognitive challenge is added, and the effects of repeated head traumas.

Martini and colleagues investigated gait stepping characteristics under both single and dual task as well as unobstructed and obstacle avoidance conditions between individuals with a self-reported history of concussions and those who denied concussion history (Martini et al., 2011). Despite an average of over 6 years having passed since the experimental groups last known/reported concussion, they still demonstrated impaired gait performance; notably, decreased gait velocity and time in single support as well as increased time in double support across tasks. The authors conclude that this represents a self-selected conservative gait pattern with a task-response relationship whereby more challenging tasks were associated with a more conservative gait strategy. This initial study was limited by small groups (68 total participants) and self-reported concussion history, the findings provide initial evidence supporting the hypothesis of delayed or incomplete neurological recovery following a concussion.

Several larger studies investigating a diverse array of post-concussion complications have provided limited gait assessment (Vanderploeg, Curtiss, Luis, & Salazar, 2007). A large scale, over 4,000 participants, cross sectional study of individuals who suffered motor vehicle accident related concussions identified a three-fold association between concussion and impaired tandem gait. This impairment was, not surprisingly, also associated with self-reported attentional deficits. Long term gait and balance impairments were recently investigated with two follow-up dates, one year and four years post-concussion, and identified approximately 1/3rd of the participants self-reported balance and gait difficulties up to 4 years post-injury (Kleffelgaard, Roe, Soberg, & Bergland, 2012). Interestingly, a key finding was that once these post-concussion symptoms present, they appear to persist over time. However, these impairments were fairly mild as none of the post-concussion participants scored below the cutoff (19) of the dynamic gait index. These studies, while limited in scope or biomechanical assessment provide a foundation for future studies to continue to investigate long term gait impairments in individuals with a concussion history.

Future Directions

Recently, considerable attention has been given to the role of repeated subconcussive impacts and their role in short- and long-term pathologies (Dashnaw, Petraglia, & Bailes, 2012). Two case studies in the mid-2000's first introduced the possibility of chronic traumatic encephalopathy (CTE), a progressive tauopathy, as a consequence of repeated head trauma associated with long term football participation (Omalu et al., 2006; Omalu et al., 2005). These case studies were followed up by larger case reports of up to 68 CTE cases in former football and hockey players as well as military veterans (A. C. McKee, et al., 2012). Further, several studies have identified altered fMRI in asymptomatic high school football players also supporting the argument of subconcussive impacts (Breedlove et al., 2012; Talavage et al., 2010). It is worth noting, however, that several other studies have failed to identify differences associated with subconcussive impacts (Gysland et al., 2012; McAllister et al., 2012; Miller, Adamson, Pink, & Sweet, 2007). From a gait perspective, one study has indirectly investigated the potential role of subconcussive blows by comparing four groups– two contact sport athlete groups and two normal healthy groups with each group subdivided into an acute concussion and no concussion history category (Parker, et al., 2008). Surprisingly, the "normal athlete" group (individuals who were NCAA division I student-athletes or club sport athletes, sports not specified in the study beyond high or low velocity impacts, without a history of concussions within the previous year) had noted reductions in gait velocity (1.21 m/s) which was actually slower than the acutely concussed participants. Overall, a continuum of performance appeared to exist whereby the normal athletes consistently performed worse than the normal non-athletes, and only marginally better than acutely concussed athletes, with increased frontal plane COM displacement and velocity as well as reduced sagittal plane separation of the COP and COM. This provocative study should be expanded with larger groups sizes, a wide array of tasks, more clearly defined sport and position definitions, and, ideally, measurements of head impact characteristics to more clearly elucidate the interaction between head impacts, subconcussive blows, and gait characteristics. If expanded upon and repeated, this could provide an early indication of the development of potential long-term pathologies.

The field of gait analysis following a sports-related concussion remains largely in its infancy when compared to the wealth of literature available on aging and diseased populations where it is not uncommon to find hundreds, if not thousands, of publications. Thus, there are numerous areas of gait analysis which can be explored to further our understanding of postural control and balance mechanisms following a concussion. Specifically, without the intention of providing a complete list, the most substantial need is to compare post-concussion gait performance through a within-subjects model whereby the injured individual is compared to their own baseline values. This could allow investigators to further the understanding of post-concussion compensatory mechanisms and or strategies being utilized by the individual as well as provide for individualized identification of post-concussion recovery. While this would be time consuming, and perhaps impractical, targeted baseline testing, perhaps of sub-groups of incoming student athletes, could be plausible. Additional areas of study, which could further the understanding of post-concussion postural control, include utilizing transitional movements (e.g., gait initiation or termination, sit-to-walk), gait performed at varying speeds to both control for changes simply due to velocity or further challenge an individual by intentionally increasing the velocity, analysis of joint or segment kinematics and kinetic assessments are warranted to further understand the altered strategy. The majority of studies have utilized working memory tasks based on the mini-mental status exam; (Bell & Hall, 1977) however, more sophisticated challenges such as Stroop color tests or similar could enhance the challenge and further stress potential compensatory strategies being utilized. Finally, the issue of cognitive rest is often raised as an imperative component of concussion management, despite a lack of empirical support,

(Moser, Glatts, & Schatz, 2012) and gait may be an interesting tool to assess the efficacy of cognitive rest and recovery.

References

Al-Yahya, E., Dawes, H., Smith, L., Dennis, A., Howells, K., & Cockburn, J. (2011). Cognitive motor interference while walking: A systematic review and meta-analysis. *Neuroscience and Biobehavioral Reviews, 35*(3), 715-728. doi: 10.1016/j. neubiorev.2010.08.008

Bell, R., & Hall, R. C. W. (1977). Mental Status Examination. *American Family Physician, 16*(5), 145-152.

Bey, T., & Ostick, B. (2009). Second impact syndrome. *The western journal of emergency medicine, 10*(1), 6-10.

Bloem, B. R., Valkenburg, V. V., Slabbekoorn, M., & Willemsen, M. D. (2001). The Multiple Tasks Test - Development and normal strategies. [Article]. *Gait & Posture, 14*(3), 191-202.

Boden, B. P., Tacchetti, R. L., Cantu, R. C., Knowles, S. B., & Mueller, F. O. (2007). Catastrophic head injuries in high school and college football players. *American Journal of Sports Medicine, 35*(7), 1075-1081. doi: 10.1177/0363546507299239

Breedlove, E. L., Robinson, M., Talavage, T. M., Morigaki, K. E., Yoruk, U., O'Keefe, K., Nauman, E. A. (2012). Biomechanical correlates of symptomatic and asymptomatic neurophysiological impairment in high school football. *Journal of Biomechanics, 45*(7), 1265-1272. doi: 10.1016/j.jbiomech.2012.01.034

Broglio, S. P., Macciocchi, S. N., & Ferrara, M. S. (2007). Sensitivity of the concussion assessment battery. *Neurosurgery, 60*(6), 1050-1057. doi: 10.1227/01.neu.0000 255479. 90999.c0

Broglio, S. P., Sosnoff, J. J., Rosengren, K. S., & McShane, K. (2009). A Comparison of Balance Performance: Computerized Dynamic Posturography and a Random Motion Platform. *Archives of Physical Medicine and Rehabilitation, 90*(1), 145-150. doi: 10.1016/j.apmr.2008.06.025

Broglio, S. P., Tomporowski, P. D., & Ferrara, M. S. (2005). Balance performance with a cognitive task: A dual-task testing paradigm. *Medicine and Science in Sports and Exercise, 37*(4), 689-695. doi: 10.1249/01.mss.0000159019.14919.09

Cantu, R. C. (1998). Second-impact syndrome. *Clinics in Sports Medicine, 17*(1), 37-+. doi: 10.1016/s0278-5919(05)70059-4

Catena, R. D., van Donkelaar, P., & Chou, L.-S. (2007). Altered balance control following concussion is better detected with an attention test during gait. *Gait & Posture, 25*(3), 406-411. doi: 10.1016/j.gaitpost.2006.05.006

Catena, R. D., van Donkelaar, P., & Chou, L.-S. (2009). Different gait tasks distinguish immediate vs. long-term effects of concussion on balance control. *Journal of Neuroengineering and Rehabilitation, 6*. doi: 25.10.1186/1743-0003-6-25

Catena, R. D., van Donkelaar, P., & Chou, L.-S. (2011). The effects of attention capacity on dynamic balance control following concussion. *Journal of Neuroengineering and Rehabilitation, 8*. doi: 810.1186/1743-0003-8-8

Catena, R. D., van Donkelaar, P., & Chou, L. S. (2007). Cognitive task effects on gait stability following concussion. *Exp Brain Res, 176*(1), 23-31. doi: 10.1007/s00221-006-0596-2

Catena, R. D., van Donkelaar, P., Halterman, C. I., & Chou, L. S. (2009). Spatial orientation of attention and obstacle avoidance following concussion. *Exp. Brain Res., 194*(1), 67-77. doi: 10.1007/s00221-008-1669-1

Cavanaugh, J. T., Guskiewicz, K. M., Giuliani, C., Marshall, S., Mercer, V. S., & Stergiou, N. (2006). Recovery of postural control after cerebral concussion: New insights using approximate entropy. [Article]. *Journal of Athletic Training, 41*(3), 305-313.

Cavanaugh, J. T., Guskiewicz, K. M., & Stergiou, N. (2004). *Detecting altered postural control after cerebral concussion in athletes without postural instability.*

Cavanaugh, J. T., Guskiewicz, K. M., & Stergiou, N. (2005). A nonlinear dynamic approach for evaluating postural control: new directions for the management of sport-related cerebral concussion. *Sports Med., 35*(11), 935-950. doi: 35112 [pii]

Chen, H. C., Schultz, A. B., AshtonMiller, J. A., Giordani, B., Alexander, N. B., & Guire, K. E. (1996). Stepping over obstacles: Dividing attention impairs performance of old more than young adults. [Article]. *Journals of Gerontology Series a-Biological Sciences and Medical Sciences, 51*(3), M116-M122.

Chen, J. K., Johnston, K. M., Frey, S., Petrides, M., Worsley, K., & Ptito, A. (2004). Functional abnormalities in symptomatic concussed athletes: an MRI study. *Neuroimage, 22*(1), 68-82. doi: 10.1016/j.neuroimage.2003.12.032

Chou, L.-S., Kaufman, K. R., Brey, R. H., & Draganich, L. F. (2001). Motion of the whole body's center of mass when stepping over obstacles of different heights. *Gait & Posture, 13*(1), 17-26.

Chou, L. S., Kaufman, K. R., Walker-Rabatin, A. E., Brey, R. H., & Basford, J. R. (2004). Dynamic instability during obstacle crossing following traumatic brain injury. *Gait & Posture, 20*(3), 245-254. doi: 10.1016/j.gaitpost.2003.09.007

Collins, M. W., Lovell, M. R., Iverson, G. L., Cantu, R. C., Maroon, J. C., & Field, M. (2002). Cumulative effects of concussion in high school athletes. *Neurosurgery, 51*(5), 1175-1179. doi: 10.1227/01.neu.0000031572.99927.93

Daneshvar, D. H., Riley, D. O., Nowinski, C. J., McKee, A. C., Stern, R. A., & Cantu, R. C. (2011). Long-Term Consequences: Effects on Normal Development Profile After Concussion. *Physical Medicine and Rehabilitation Clinics of North America, 22*(4), 683-+. doi: 10.1016/j.pmr.2011.08.009

Dashnaw, M. L., Petraglia, A. L., & Bailes, J. E. (2012). An overview of the basic science of concussion and subconcussion: where we are and where we are going. *Neurosurgical Focus, 33*(6), E5. doi: 10.3171/2012.10.focus12284

Fait, P., McFadyen, B. J., Swaine, B., & Cantin, J. F. (2009). Alterations to locomotor navigation in a complex environment at 7 and 30 days following a concussion in an elite athlete. *Brain Injury, 23*(4), 362-369. doi: Pii 90931067810.1080/02699050902788485

Finnoff, J. T., Peterson, V. J., Hollman, J. H., & Smith, J. (2009). Intrarater and Interrater Reliability of the Balance Error Scoring System (BESS). *Pm&R, 1*(1), 50-54. doi: 10.1016/j.pmrj.2008.06.002

Fox, Z. G., Mihalik, J. P., Blackburn, J. T., Battaglini, C. L., & Guskiewicz, K. M. (2008). Return of postural control to baseline after anaerobic and aerobic exercise protocols. *Journal of Athletic Training, 43*(5), 456-463. doi: 10.4085/1062-6050-43.5.456

Galna, B., Peters, A., Murphy, A. T., & Morris, M. E. (2009). Obstacle crossing deficits in older adults: A systematic review. *Gait & Posture, 30*(3), 270-275. doi: 10.1016/j.gaitpost.2009.05.022

Gao, J., Hu, J., Buckley, T., White, K., & Hass, C. (2011). Shannon and Renyi Entropies to Classify Effects of Mild Traumatic Brain Injury on Postural Sway. *PLoS One, 6*(9). doi: e2444610.1371/journal.pone.0024446

Giza, C. C., & Hovda, D. A. (2001). The neurometabolic cascade of concussion. *Journal of Athletic Training, 36*(3), 228-235.

Guilmette, T. J., & Paglia, M. F. (2004). The public's misconceptions about traumatic brain injury: a follow up survey. *Archives of Clinical Neuropsychology, 19*(2), 183-189. doi: 10.1016/s0887-6177(03)00025-8

Guskiewicz, K. M. (2011). Balance Assessment in the Management of Sport-Related Concussion. *Clinics in Sports Medicine, 30*(1), 89-+. doi: 10.1016/j.csm.2010.09.004

Guskiewicz, K. M., Bruce, S. L., Cantu, R. C., Ferrara, M. S., Kelly, J. P., McCrea, M., . . . McLeod, T. C. V. (2004). National Athletic Trainers' Association position statement: Management of sport-related concussion. *Journal of Athletic Training, 39*(3), 280-297.

Guskiewicz, K. M., Marshall, S. W., Bailes, J., McCrea, M., Cantu, R. C., Randolph, C., & Jordan, B. D. (2005). Association between recurrent concussion and late-life cognitive impairment in retired professional football players. *Neurosurgery, 57*(4), 719-724. doi: 10.1227/01.neu.0000175725.75780.dd

Guskiewicz, K. M., Marshall, S. W., Bailes, J., McCrea, M., Harding, H. P., Jr., Matthews, A., Cantu, R. C. (2007). Recurrent concussion and risk of depression in retired professional football players. *Medicine and Science in Sports and Exercise, 39*(6), 903-909. doi: 10.1249/mss.0b013e3180383da5

Guskiewicz, K. M., McCrea, M., Marshall, S. W., Cantu, R. C., Randolph, C., Barr, W., . . . Kelly, J. P. (2003). Cumulative effects associated with recurrent concussion in collegiate football players - The NCAA Concussion Study. *Jama-Journal of the American Medical Association, 290*(19), 2549-2555. doi: 10.1001/jama.290.19.2549

Guskiewicz, K. M., Perrin, D. H., & Gansneder, B. M. (1996). Effect of mild head injury on postural stability in athletes. *Journal of Athletic Training, 31*(4), 300-306.

Guskiewicz, K. M., Ross, S. E., & Marshall, S. W. (2001). Postural stability and neuropsychological deficits after concussion in collegiate athletes. *Journal of Athletic Training, 36*(3), 263-273.

Guskiewicz, K. M., Weaver, N. L., Padua, D. A., & Garrett, W. E. (2000). Epidemiology of concussion in collegiate and high school football players. *American Journal of Sports Medicine, 28*(5), 643-650.

Gysland, S. M., Mihalik, J. P., Register-Mihalik, J. K., Trulock, S. C., Shields, E. W., & Guskiewicz, K. M. (2012). The Relationship Between Subconcussive Impacts and Concussion History on Clinical Measures of Neurologic Function in Collegiate Football Players. *Annals of Biomedical Engineering, 40*(1), 14-22. doi: 10.1007/s10439-011-0421-3

Hass, C. J., Gregor, R. J., Waddell, D. E., Oliver, A., Smith, D. W., Fleming, R. P., & Wolf, S. L. (2004). The influence of Tai Chi training on the center of pressure trajectory during gait initiation in older adults. *Archives of Physical Medicine and Rehabilitation, 85*(10), 1593-1598. doi: 10.1016/j.apmr.2004.01.020

Hass, C. J., Waddell, D. E., Fleming, R. P., Juncos, J. L., & Gregor, R. J. (2005). Gait initiation and dynamic balance control in Parkinson's disease. *Archives of Physical Medicine and Rehabilitation, 86*(11), 2172-2176. doi: 10.1016/j.apmr.2005.05.013

Hass, C. J., Waddell, D. E., Wolf, S. L., Juncos, J. L., & Gregor, R. J. (2008). Gait initiation in older adults with postural instability. [Article]. *Clinical Biomechanics, 23*(6), 743-753. doi: 10.1016/j.clinbiomech.2008.02.012

Hass, C. J., Wikstrom, E., Fournier, K., Inamdar, A., & Bishop, M. (2008). *Locomotor Initiation: Influence of Chronic Ankle Instability.* Paper presented at the North American Congress on Biomechanics Ann Arbor, MI.

Hausdorff, J. M., Schweiger, A., Herman, T., Yogev-Seligmann, G., & Giladi, N. (2008). Dual-Task Decrements in Gait: Contributing Factors Among Healthy Older Adults. [Article]. *Journals of Gerontology Series a-Biological Sciences and Medical Sciences, 63*(12), 1335-1343.

Herring, S. A., Cantu, R. C., Guskiewicz, K. M., Putukian, M., Kibler, W. B., Bergfeld, J. A., . . . American College of Sports, M. (2011). Concussion (mild traumatic brain injury) and the team physician: a consensus statement--2011 update. *Medicine and Science in Sports and Exercise, 43*(12), 2412-2422.

Hovda, D. A., Yoshino, A., Kawamata, T., Katayama, Y., & Becker, D. P. (1991). Diffuse prolonged depression of cerebral oxidative-metabolism following concussive brain injury in the rat - a cytochrome-oxidase histochemistry study. *Brain Research, 567*(1), 1-10. doi: 10.1016/0006-8993(91)91429-5

Hoxie, R. E., Rubenstein, L. Z., Hoenig, H., & Gallagher, B. R. (1994). The older pedestrian. *J. Am. Geriatr Soc., 42*(4), 444-450.

Iverson, G. L., Gaetz, M., Lovell, M. R., & Collins, M. W. (2004). Cumulative effects of concussion in amateur athletes. [Article]. *Brain Injury, 18*(5), 433-443. doi: 10.1080/02699050310001617352

Kelly, J. P., & Rosenberg, J. H. (1997). Diagnosis and management of concussion in sports. *Neurology, 48*(3), 575-580.

Kerr, Z. Y., Marshall, S. W., & Guskiewicz, K. M. (2012). Reliability of Concussion History in Former Professional Football Players. *Medicine and Science in Sports and Exercise, 44*(3), 377-382. doi: 10.1249/MSS.0b013e31823240f2

Kleffelgaard, I., Roe, C., Soberg, H. L., & Bergland, A. (2012). Associations among self-reported balance problems, post-concussion symptoms and performance-based tests: a longitudinal follow-up study. *Disability and Rehabilitation, 34*(9), 788-794. doi: 10.3109/09638288.2011.619624

Kovacs, C. R. (2005). Age-related changes in gait and obstacle avoidance capabilities in older adults: A review. *Journal of Applied Gerontology, 24*(1), 21-34. doi: 10.1177/0733464804271279

Langlois, J. A., Keyl, P. M., Guralnik, J. M., Foley, D. J., Marottoli, R. A., & Wallace, R. B. (1997). Characteristics of older pedestrians who have difficulty crossing the street. *Am J Public Health, 87*(3), 393-397.

Langlois, J. A., Rutland-Brown, W., & Wald, M. M. (2006). The epidemiology and impact of traumatic brain injury - A brief overview. *Journal of Head Trauma Rehabilitation, 21*(5), 375-378.

Lee, H., Sullivan, S. J., & Schneiders, A. G. (2012). The use of the dual-task paradigm in detecting gait performance deficits following a sports-related concussion: A systematic review and meta-analysis. *J. Sci. Med. Sport*. doi: 10.1016/j.jsams.2012.03.013

Marois, R., & Ivanoff, J. (2005). Capacity limits of information processing in the brain. [Review]. *Trends in Cognitive Sciences, 9*(6), 296-305. doi: 10.1016/j.tics.2005.04.010

Martini, D. N., Sabin, M. J., DePesa, S. A., Leal, E. W., Negrete, T. N., Sosnoff, J. J., & Broglio, S. P. (2011). The Chronic Effects of Concussion on Gait. *Archives of Physical Medicine and Rehabilitation, 92*(4), 585-589. doi: 10.1016/j.apmr.2010.11.029

McAllister, T. W., Flashman, L. A., Maerlender, A., Greenwald, R. M., Beckwith, J. G., Tosteson, T. D., . . . Turco, J. H. (2012). Cognitive effects of one season of head impacts in a cohort of collegiate contact sport athletes. *Neurology, 78*(22), 1777-1784. doi: 10.1212/WNL.0b013e3182582fe7

McCrea, M., Guskiewicz, K., Randolph, C., Barr, W. B., Hammeke, T. A., Marshall, S. W., & Kelly, J. P. (2009). Effects of a symptom-free waiting period on clinical outcome and risk of reinjury after sport-related concussion. *Neurosurgery, 65*(5), 876-882. doi: 10.1227/01.neu.0000350155.89800.00

McCrea, M., Guskiewicz, K. M., Marshall, S. W., Barr, W., Randolph, C., Cantu, R. C., . . . Kelly, J. P. (2003). Acute effects and recovery time following concussion in collegiate football players - The NCAA Concussion Study. *Jama-Journal of the American Medical Association, 290*(19), 2556-2563. doi: 10.1001/jama.290.19.2556

McCrea, M., Hammeke, T., Olsen, G., Leo, P., & Guskiewicz, K. (2004). Unreported concussion in high school football players - Implications for prevention. *Clinical Journal of Sport Medicine, 14*(1), 13-17. doi: 10.1097/00042752-200401000-00003

McCrory, P. (2001). Does second impact syndrome exist? *Clinical Journal of Sport Medicine, 11*(3), 144-149. doi: 10.1097/00042752-200107000-00004

McCrory, P., Davis, G., & Makdissi, M. (2012). Second Impact Syndrome or Cerebral Swelling after Sporting Head Injury. *Current sports medicine reports, 11*(1), 21-23. doi: 10.1249/JSR.0b013e3182423bfd

McCrory, P., Meeuwisse, W., Johnston, K., Dvorak, J., Aubry, M., Molloy, M., & Cantu, R. (2009). Consensus statement on concussion in sport: the 3rd International Conference on Concussion in Sport held in Zurich, November 2008. *Journal of Athletic Training, 44*(4), 434-448. doi: 10.4085/1062-6050-44.4.434

McCrory, P., Meeuwisse, W., Johnston, K., Dvorak, J., Aubry, M., Molloy, M., & Cantu, R. (2009). Consensus statement on concussion in sport: the 3rd International Conference on Concussion in Sport held in Zurich, November 2008. *J. Athl. Train, 44*(4), 434-448.

McCrory, P. R., & Berkovic, S. F. (2000). Video analysis of acute motor and convulsive manifestations in sport-related concussion. *Neurology, 54*(7), 1488-1491.

McCrory, P. R., & Berkovic, S. F. (2001). Concussion: the history of clinical and pathophysiological concepts and misconceptions. *Neurology, 57*(12), 2283-2289.

McKee, A. C., Cantu, R. C., Nowinski, C. J., Hedley-Whyte, E. T., Gavett, B. E., Budson, A. E., Stern, R. A. (2009). Chronic Traumatic Encephalopathy in Athletes: Progressive Tauopathy After Repetitive Head Injury. *Journal of Neuropathology and Experimental Neurology, 68*(7), 709-735.

McKee, A. C., Gavett, B. E., Stern, R. A., Nowinski, C. J., Cantu, R. C., Kowall, N. W., . . . Budson, A. E. (2010). TDP-43 Proteinopathy and Motor Neuron Disease in Chronic

Traumatic Encephalopathy. *Journal of Neuropathology and Experimental Neurology, 69*(9), 918-929. doi: 10.1097/NEN.0b013e3181ee7d85

McKee, A. C., Stein, T. D., Nowinski, C. J., Stern, R. A., Daneshvar, D. H., Alvarez, V. E., . . . Cantu, R. C. (2012). The spectrum of disease in chronic traumatic encephalopathy. *Brain*. doi: 10.1093/brain/aws307

Miller, J. R., Adamson, G. J., Pink, M. M., & Sweet, J. C. (2007). Comparison of preseason, midseason, and postseason neurocognitive scores in uninjured collegiate football players. *American Journal of Sports Medicine, 35*(8), 1284-1288. doi: 10.1177/036354 6507300261

Moser, R. S., Glatts, C., & Schatz, P. (2012). Efficacy of Immediate and Delayed Cognitive and Physical Rest for Treatment of Sports-Related Concussion. *Journal of Pediatrics, 161*(5), 922-926. doi: 10.1016/j.jpeds.2012.04.012

NINDS. (2012, June 4, 2012). Traumatic Brain Injury Retrieved December 12, 2012, from http://www.ninds.nih.gov/disorders/tbi/detail_tbi.htm

Omalu, B. I., DeKosky, S. T., Hamilton, R. L., Minster, R. L., Kamboh, M. I., Shakir, A. M., & Wecht, C. H. (2006). Chronic traumatic encephalopathy in a national football league player: Part II. *Neurosurgery, 59*(5), 1086-1092. doi: 10.1227/01.neu.000024 5601.6945127

Omalu, B. I., DeKosky, S. T., Minster, R. L., Kamboh, M. I., Hamilton, R. L., & Wecht, C. H. (2005). Chronic traumatic encephalopathy in a National Football League player. *Neurosurgery, 57*(1), 128-133. doi: 10.1227/01.neu.0000163407.92769.ed

Onate, J. A., Beck, B. C., & Van Lunen, B. L. (2007). On-field testing environment and balance error scoring system performance during preseason screening of healthy collegiate baseball players. *Journal of Athletic Training, 42*(4), 446-451.

Parker, T. M., Osternig, L. R., & Chou, L.-S. (2006). Gait Stability in Athletes and Non-Athletes Following Concussion. *Medicine and Science in Sports and Exercise, 38*(5), S2-S2. doi: 10.1249/00005768-200605001-00028

Parker, T. M., Osternig, L. R., Lee, H. J., van Donkelaar, P., & Chou, L. S. (2005). The effect of divided attention on gait stability following concussion. *Clinical Biomechanics, 20*(4), 389-395. doi: 10.1016/j.clinbiomech.2004.12.004

Parker, T. M., Osternig, L. R., van Donkelaar, P., & Chou, L.-S. (2007). Recovery of cognitive and dynamic motor function following concussion. *British Journal of Sports Medicine, 41*(12), 868-873. doi: 10.1136/bjsm.2006.033761

Parker, T. M., Osternig, L. R., Van Donkelaar, P., & Chou, L. S. (2006). Gait stability following concussion. *Medicine and Science in Sports and Exercise, 38*(6), 1032-1040. doi: 10.1249/01.mss.0000222828.56982.a4

Parker, T. M., Osternig, L. R., van Donkelaar, P., & Chou, L. S. (2008). Balance control during gait in athletes and non-athletes following concussion. *Med. Eng. Phys., 30*(8), 959-967. doi: 10.1016/j.medengphy.2007.12.006

Pashler, H. (1994). Dual-Task Interference in Simple Tasks - Data and Theory. [Review]. *Psychological Bulletin, 116*(2), 220-244.

Perry, J. (1992). *Gait Analysis: Normal and Pathological Function* (1st ed.). Thorofare, NJ: SLACK Inc.

Plummer-D'Amato, P., Altmann, L. J. P., Saracino, D., Fox, E., Behrman, A. L., & Marsiske, M. (2008). Interactions between cognitive tasks and gait after stroke: A dual task study. *Gait & Posture, 27*(4), 683-688. doi: 10.1016/j.gaitpost.2007.09.001

Purser, J. L., Weinberger, M., Cohen, H. J., Pieper, C. F., Morey, M. C., Li, T., . . . Lapuerta, P. (2005). Walking speed predicts health status and hospital costs for frail elderly male veterans. *J. Rehabil. Res. Dev., 42*(4), 535-546.

Resch, J. E., May, B., Tomporowski, P. D., & Ferrara, M. S. (2011). Balance Performance With a Cognitive Task: A Continuation of the Dual-Task Testing Paradigm. *Journal of Athletic Training, 46*(2), 170-175. doi: 10.4085/1062-6050-46.2.170

Riemann, B. L., Guskiewicz, K. M., & Shields, E. W. (1999). Relationship between clinical and forceplate measures of postural stability. *Journal of Sport Rehabilitation, 8*(2), 71-82.

Sharikadze, M., Cong, D. K., Staude, G., Deubel, H., & Wolf, W. (2009). Dual-tasking: Is manual tapping independent of concurrently executed saccades? *Brain Research, 1283*, 41-49. doi: 10.1016/j.brainres.2009.05.065

Shaw, N. A. (2002). Neurophysiology of concussion. *Progress in Neurobiology, 67*(4), 281-344. doi: Pii 0301-0082(02)00018-710.1016/s0301-0082(02)00018-7

Shumway-Cook, A., Woollacott, M., Kerns, K. A., & Baldwin, M. (1997). The effects of two types of cognitive tasks on postural stability in older adults with and without a history of falls. *J. Gerontol. A. Biol. Sci. Med. Sci., 52*(4), M232-240.

Slobounov, S., Cao, C., Sebastianelli, W., Slobounov, E., & Newell, K. (2008). Residual deficits from concussion as revealed by virtual time-to-contact measures of postural stability. *Clin. Neurophysiol., 119*(2), 281-289. doi: 10.1016/j.clinph.2007.10.006

Slobounov, S., Tutwiler, R., Sebastianelli, W., & Slobounov, E. (2006). Alteration of postural responses to visual field motion in mild traumatic brain injury. *Neurosurgery, 59*(1), 134-139; discussion 134-139. doi: 10.1227/01.neu.0000219197.33182.3f

Stegemoeller, E. L., Buckley, T. A., Pitsikoulis, C., Barthelemy, E., Roemmich, R., & Hass, C. J. (2012). Postural Instability and Gait Impairment During Obstacle Crossing in Parkinson's Disease. *Archives of Physical Medicine and Rehabilitation, 93*(4), 703-709. doi: 10.1016/j.apmr.2011.11.004

Susco, T. M., McLeod, T. C. V., Gansneder, B. M., & Shultz, S. J. (2004). Balance recovers within 20 minutes after exertion as measured by the balance error scoring system. *Journal of Athletic Training, 39*(3), 241-246.

Talavage, T. M., Nauman, E., Breedlove, E. L., Yoruk, U., Dye, A. E., Morigaki, K., . . . Leverenz, L. J. (2010). Functionally-Detected Cognitive Impairment in High School Football Players Without Clinically-Diagnosed Concussion. *J. Neurotrauma*. doi: 10.1089/neu.2010.1512

Teel, E. F., Register-Mihalik, J. K., Troy Blackburn, J., & Guskiewicz, K. M. (2012). Balance and cognitive performance during a dual-task: Preliminary implications for use in concussion assessment. *J. Sci. Med. Sport*. doi: 10.1016/j.jsams.2012.09.007

Valovich McLeod, T. C., Perrin, D. H., Guskiewicz, K. M., Shultz, S. J., Diamond, R., & Gansneder, B. M. (2004). Serial administration of clinical concussion assessments and learning effects in healthy young athletes. *Clinical journal of sport medicine : official journal of the Canadian Academy of Sport Medicine, 14*(5), 287-295. doi: 10.1097/00042752-200409000-00007

Valovich, T. C., Perrin, D. H., & Gansneder, B. M. (2003). Repeat administration elicits a practice effect with the balance error scoring system but not with the standardized assessment of concussion in high school athletes. *Journal of Athletic Training, 38*(1), 51-56.

van Donkelaar, P., Osternig, L., & Chou, L. S. (2006). Attentional and biomechanical deficits interact after mild traumatic brain injury. *Exercise and Sport Sciences Reviews, 34*(2), 77-82. doi: 10.1249/00003677-200604000-00007

Vanderploeg, R. D., Curtiss, G., Luis, C. A., & Salazar, A. M. (2007). Long-term morbidities following self-reported mild traumatic brain injury. *Journal of Clinical and Experimental Neuropsychology, 29*(6), 585-598. doi: 10.1080/13803390600826587

Wikstrom, E. A., Bishop, M. D., Inamdar, A. D., & Hass, C. J. (2010). Gait termination control strategies are altered in chronic ankle instability subjects. *Med Sci Sports Exerc, 42*(1), 197-205. doi: 10.1249/MSS.0b013e3181ad1e2f [doi]

Wikstrom, E. A., & Hass, C. J. (2012). Gait termination strategies differ between those with and without ankle instability. *Clinical biomechanics (Bristol, Avon), 27*(6), 619-624.

Wikstrom, E. A., Tapia-Lovler, T. G., Munkasy, B. A., & Buckley, T. A. (2012). *Concussion Alters Gait Termination Strategies.* Paper presented at the American Society of Biomechanics, Gainesville, FL.

Wilkins, J. C., McLeod, T. C. V., Perrin, D. H., & Gansneder, B. M. (2004). Performance on the balance error scoring system decreases after fatigue. *Journal of Athletic Training, 39*(2), 156-161.

Woollacott, M., & Shumway-Cook, A. (2002). Attention and the control of posture and gait: a review of an emerging area of research. *Gait Posture, 16*(1), 1-14. doi: S0966636201001564 [pii]

Zemper, E. D. (2003). Two-year prospective study of relative risk of a second cerebral concussion. *American Journal of Physical Medicine & Rehabilitation, 82*(9), 653-659. doi: 10.1097/01.phm.0000083666.74494.ba

Zhang, K., Johnson, B., Pennell, D., Ray, W., Sebastianelli, W., & Slobounov, S. (2010). Are functional deficits in concussed individuals consistent with white matter structural alterations: combined FMRI & DTI study. *Exp. Brain Res., 204*(1), 57-70. doi: 10.1007/s00221-010-2294-3.

In: Gait Biometrics
Editors: Li Li and Matthew Holmes
ISBN: 978-1-62948-381-8

Chapter IX

Biomechanical and Neuromuscular Control Factors in Determining Gait Differences in Individuals with and without Lower Back Pain

Michael W. Olson*
Southern Illinois University, Carbondale, IL, US

Abstract

The etiology of low back pain is multifaceted and not clearly understood. However, the importance of determining the significance of this symptom is paramount as nearly 80% of people during their lifetime will possess either an acute or chronic case of low back pain. As such, activities of daily living are greatly influenced by pain symptoms. The walking gait parameters of individuals afflicted with low back pain is reported to be significantly different than that of healthy controls. Similarly, biomechanical variables analyzed during running gait provide further evidence of differential movement coordination between individuals with and without low back pain. From a biomechanical perspective, the movement patterns and the forces acting upon the body are key factors in providing meaningful information regarding the significance of the influence of low back pain during the gait cycle. Further, the motor control parameters, particularly the muscle activation patterns, during the gait cycle indicate compensatory and protective mechanisms to reduce degrees of freedom in guarding against further pain. Whether these mechanisms are indicative of the system attempting to provide additional stability to the body is currently inconclusive. This review will introduce a spectrum of issues potentially related to the causation of low back pain followed by the specific modifications to biomechanical and motor control measures between healthy individuals and those with low back pain during walking and running.

* Email: mwolson@siu.edu.

Introduction

The musculoskeletal system provides humans the ability to uniquely interact with the external environment. In daily activities of living we are able to support the load of our body, maintain posture, lift external loads, perform locomotion, and many other necessary tasks. Likewise, the control of our moving limbs is regulated by the neurological information highways afforded us through the central and peripheral nervous systems in order to determine the proper responses to environmental stimuli. Human bipedal locomotion affords us a distinctive mode of travel that most other species are not capable of performing. From an evolutionary perspective, our cranial vault, spine, and pelvis have been structurally adapted to walking on two legs (Haeusler et al., 2011; Niemitz, 2010; Preuschoft, 2004). The results of these evolutionary adaptations were increased lordosis of the lumbar and cervical spine levels in order to increase force dissipation, and direct the center of mass (CM) over the base of support (Haeusler et al., 2011). The positive attributes of being bipedal affords humans the ability to reach for objects much higher than the ground level, increase the area over which we can visualize by increasing our vertical height, and enables our hands to perform tasks independent of our locomotor function (Haeusler et al., 2011; Niemitz, 2010). Conversely, there are negative attributes that are also present due to our erect stature. First and foremost, the propensity of the low back region (lumbar spine) to incur great loading can adversely influence the health of the vertebrae and the soft tissues of the system. In gait analysis, the inverted pendulum is used to describe the pure mechanical properties of the human walking mode; similarly, the head, arms, and trunk (HAT) are inverted pendulums with respect to the pelvis (MacKinnon & Winter, 1993; Winter et al., 1993). This requires the mass of the HAT to act upon the lumbar spine as this structure is positioned above the sacrum and pelvic bones. To support such a disbursed load requires guywires in the form of muscles and ligaments to sustain the position of the HAT (Crisco & Panjabi, 1991; Wilke et al., 1995). Not only does the weight of the HAT influence the loading of the lumbar spine, but the activation of the muscles (Marras & Sommerich 1991a, 1991b; McGill & Norman, 1986) and loading schemes (flexion/extension, bending, torsion, shearing) in combination (Frei et al., 2002; Wang et al., 2000) provide additional loading to the vertebrae and intervertebral discs. The trunk muscles, in particular, become increasingly important as the difficulty of the movement task increases (Kavcic et al., 2004). Locomotion requires the trunk muscles to actively control the positioning of the mass in order to successfully move within the external environment. This task is further complicated as the spine must be able to attenuate the ground reaction forces transmitted up the leg to the pelvis and spine (Voloshin & Wosk, 1982). If pain is present in the low back region it may be difficult for the spine to properly attenuate the force transmission from the ground reaction forces during gait. The musculoskeletal and neuromuscular systems attempt to compensate to reduce spinal loading when low back pain is present. However, when compared with healthy age-matched controls, those with low back are observed to have greater knee joint stiffness, thus exasperating the transfer of ground reaction forces to the spine during running (Hamill et al., 2009).

In healthy individuals, the daily loads imparted to the spine are sufficient to stimulate the neuromuscular system response. Cumulative loading schemes (Kerr et al., 2001; Norman et al., 1998; van Dieën et al., 1999), magnitude of the loading (Coenen et al., 2012), and the combination of loading schemes (Wisleder et al,. 2001) during daily activities are a handful of factors that will influence whether the musculoskeletal system will be able to support and sustain the imparted loads. The task of walking is not documented to be a cause of low back pain or injury, and is in fact, considered to be beneficial to the generational span of individuals in regards to cardiovascular health, and its relatively low loading even at higher walking speeds (Callahan et al., 1999). However, individuals possessing low back pain can find walking to be very problematic. Physical signs of trauma to the low back tissues and, in many cases, undetermined origins can result in modified load distribution during the gait cycle leading to limited ranges of motion for both the trunk and lower extremities (Simmonds et al,. 2012; van den Hoorn et al., 2012; Vogt et al., 2001). This literature review intends to highlight research involving low back pain and the ramifications of this affliction to bipedal gait in order to provide an understanding of the link between low back pain and locomotion. Further, biomechanical, neuromuscular, and motor control mechanisms will be evaluated and explored in determining the key attributes and difference between symptomatic (pain) and asymptomatic (pain-free) individuals during locomotion.

In this chapter, the relationship between low back pain and human bipedal locomotion will be presented from a systematic perspective. Since the incidence of low back pain is believed to originate from many different causes, and is still difficult to accurately determine, idiopathic low back pain (i.e., no physical signs of trauma) will be the primary condition addressed. A discussion of the symptomatic causes and ramification of the pain incurred to the system's control and stability will be presented. The issues presented are inclusive of kinetic and kinematics parameterizations as well as motor control factors overlying the biomechanical constraints and variability during the motor tasks. Further, acute and chronic changes to low back pain will be addressed and compared with asymptomatic individuals with clinical reports of how biomechanical and motor control variables related to gait change during recovery from pain. Key word searches were performed using the terms "etiology and epidemiology of low back pain", "low back pain and gait", "gait adaptations to pain", "locomotion control" and "biomechanical variables of gait" for pertinent literature using PubMed and Web of Knowledge (Science) databases as the majority of the data related to biomechanical research regarding gait is within these repositories.

Review

Epidemiology of Low Back Pain

The incidence of low back pain and injury is a major concern in industrialized nations. The United States Department of Labor reports 43 incidence of low back injury or pain per every 10,000 worker from 2003-2006, the highest rate among any part of the body (USLD, 2007). These incidences require approximately eight days away from the workforce, costing billions of dollars to the individual, the employer, and insurance companies (Maetzel & Li, 2002; Maniadakis & Gray, 2000). Interestingly, the number of musculoskeletal disorders has steadily decreased from 34% to 30.2% of all work-related cases from 2002 to 2006 (USDL, 2007). Still, this is a major concern to the general population as at least 80% of us will possess either an acute or chronic case of low back pain in our lifetime (Bering-Sørenson, 1983, Frymoyer et al., 1980; Svensson et al., 1989). Further, the recurrence of low back pain is an additional concern, especially if initial medical treatments were not effective at alleviating the source of the pain (Bering-Sørenson, 1983; Jorgensen, 2007). We currently do not fully understand all the potential risk factors leading to first occurrences or additional bouts of low back pain (Oleske et al., 2006; Woolf & Pfleger, 2003). A number of risk factors have been suggested that may predispose individuals to a higher propensity of low back pain. Reports of low back pain are shown from individuals in occupations such as manual material handling (Oleske et al., 2004; 2006) and medical professions (Adams et al., 1999; Jensen et al., 2012), while factors such pregnancy (Malmqvist et al., 2012; Ostgaard et al., 1991), obesity (Cimolin et al., 2011; Fanuele et al., 2002) and smoking/cardiovascular health (Frymoyer et al., 1980; Svensson et al., 1983) are also related to these incidents.

Specific areas of injury and concentrated stresses are documented in the literature and contribute to pain, whether the etiology is known or the pain symptoms are idiopathic. The radiographic and magnetic resonance images of disc protrusion, disc degeneration, and vertebral damage can be readily apparent and useful in explaining low back pain (Adams & Dolan, 2005). Many times, however, these imaging techniques are not helpful in successfully determining root cause of the pain source. Damage to muscles and other connective tissues due to extraneous efforts during high magnitude loading (Coenen et al., 2012) or increased loading frequency (de Looze et al,. 1996) are potential causes for low back pain and are usually acute and are alleviate relatively quickly. As the muscles are used to produce high magnitude forces to overcome or balance external loads imposed upon the spine, these forces also act to increase the loading of the intervertebral discs and articulating processes. As a result of these forces the disc responds by reducing height as fluids are expelled when the disc stresses increase. The modified biomechanical behavior of the discs may result in stiffening of the tissues providing for greater rigidity, leading to a modified distribution of loading of not only the disc, but the neural arch as well (Pollintine et al., 2004a: 2004b). These biomechanical behavior changes also modify the response of mechanoreceptors (Ruffini ending, Pacininan corpuscles) embedded within the discs and surrounding connective tissues (ligaments, fascia) (Solomonow, 2006; Yahia & Newman, 1991). Further, afferent information from the embedded mechanoreceptors is believed to project to gamma (γ) - motorneurons which can significantly alter the response of the muscle (see Sjolander et al., 2002) (see Figure 1). Pain from these areas may be a result of the changes to the properties of the soft viscoelastic tissues which stimulate low threshold nociceptors. Additionally, synovial joint compression and shearing at the apophyseal articulations may introduce damage to the boney processes leading to low back pain (Indahl et al., 1997). Structural construction and diseases directly influencing the structure of the spine (spondylolisthesis, spondylolysis, scoliosis, degenerative disc disease, etc.) are further concerns to orthopedists, neurologist, and therapists in providing corrective measures and pain alleviation (O'Sullivan et al., 1997). Biomechanical behavior changes of the discs leading to degeneration also contribute to the compression of the spinal nerves resulting in radiating pain (Millisdotter et al., 2003; Yamashita et al,. 2002). The lumbodorsal fascia and ligamentous support at each vertebral level contribute to the support of the lumbar spine and are potential sources of low back pain as micro-damage is possible after prolonged loading (Solomonow et al., 1999; 2002). Much of the damage to the connective tissues is speculative, but there is evidence that micro damage is present as neutrophil and cytokine concentrations and other physiological markers indicate (Solomonow, 2011a; Solomonow et al., 2011b). Evidence is present to suggest biomechanical behavior of the viscoelastic passive tissues, including disc and fascia/ligament, is influenced by anterior bending of the lumbar spine (Adams et al., 1994). The mechanical creep or tension-relaxation developed in the ligamentous tissues reduces the force transmission ability of the network of interconnected tissues (Olson, 2011) and may reduce the mechanical stability of the spine (McGill & Brown, 1992). The loading schemes specific to the lumbar spine are expanded upon in the next section and will highlight the intricacies involving the potential etiologies of low back pain.

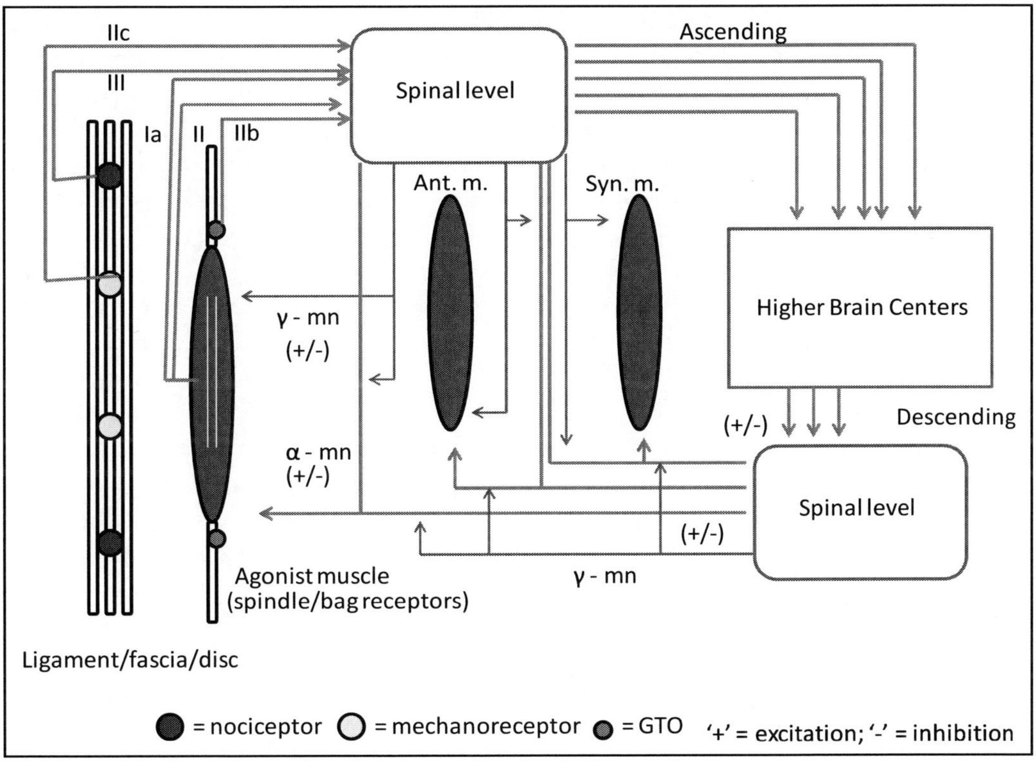

Figure 1. Mechanistic pathway for mechanoreceptor and nociceptive input to the central nervous system. Pathways indicate short-loop (spinal level) and long-loop (higher brain center) responses to afferent input. Efferent (α - mn) pathways from the CNS are provided for antagonist (Ant. m.) and synergist (Syn. m.) muscles. The gamma motorneuron (γ - mn) pathways are included to illustrate their influence on α – mn pathways and muscle spindles.

Loading Schemes Influencing the Lumbar Spine

The lumbar spine allows humans the ability to maintain an upright erect posture during many activities throughout the day. However, the erect stature of the spine is not always maintained as we manipulate objects within our environment for leisure, sport, occupation, or everyday tasks. While we are standing the spine is primarily loaded in compression with the nucleus pulposus being stressed between in the intervertebral bodies of the adjacent vertebrae, whereas the concentric layers of anulus fibrosis resist the pushing of the nucleus pulposus material away for the disc center. The lumbar spine is anteriorly concave (lordodic) so there is no pure compressive stress acting upon these vertebrae or the intervertebral discs. The lumbar lordosis results in greater shear stresses, bending, and potential uneven loading of the intervertebral discs, in comparison to pure compressive loading (Keller et al., 2005; Shirazi-Adl & Parnianpour, 1999). In vitro studies have reported the deviation of the nucleus pulposus in opposition to the region of loading during bending movements in healthy specimen (Tsantrizos et al., 2005). During flexion-extension and lateral bending movements the instantaneous axis is also shown to move towards the direction of the movement (Rousseau et al., 2006) indicating the dynamic nature of the viscoelastic tissues. Additionally, the inferior and superior articulating processes which form the articular/zygapophysial joints between the adjacent vertebrae will be stressed when the spine is moved into bending, torsion, and side bending (Bogduk, 1999). These articular joints provide supplementary support to the vertebral body and intervertebral discs (Adams & Dolan, 2005) and act to resist extraneous vertebral movement.

The spine relies greatly on the ligamentous support, as well as the muscles, to maintain positioning of the trunk. Anterior and posterior longitudinal ligaments are constructed of thick collagenous fibers of connective tissues which run anteriorly and posteriorly along the lumbar spine to provide support to the vertebral bodies and discs. Smaller, but just as important, are the ligaments that are posterior to the vertebral foramen as these ligaments resist antero-posterior bending, lateral bending, and torsion (Adams & Dolan, 1996; 2005; Hindle et al., 1990). In particular, the ligamentum flavum (yellow ligament) along with the articular ligament, interspinous ligament, and supraspinous ligament act to transfer loads and passively resist movements of the spine (Chazal et al., 1985; Dumas et al., 1987; Pintar et al., 1992; Sharma et al., 1995). A relatively large thick collagenous ligament, called the iliolumbar ligament, spans from the fifth vertebrae to the sacrum and resists anterior sliding of the lumbar spine relative to the sacrum (Bogduk, 1999). This ligament becomes very important when determining the loading of the spine and corresponding movement patterns between the lumbar region and the pelvis.

The muscles of the spine provide the dynamic stability necessary for performing daily activities such as walking, lifting, and sitting. The muscles which innervate the spine are highly specialized for providing the spine the means of resisting external loading. For example, the center of mass (CM) of the body is located near the lumbar spine which results in very little muscle activity requirement to sustain the upright standing posture. However, trunk flexion or redistribution of the CM anterior relative to the spine results in a flexion moment about the intervertebral joint (discs) necessitating the activation of the trunk extensors to sustain the body's position. The internal stresses within passive ligaments and fascia are initially low as the movement begins and requires the muscles of the low back to activate and develop an extension moment to offset the body weight, allowing for the system to maintain its equilibrium. However, with additional anterior flexion of the trunk, the stresses of the ligaments and fascia increase to support the trunk load in healthy individuals (Dolan et al., 1994; Floyd & Silver, 1951, Floyd & Silver, 1955; McGill & Kippers, 1994), with a corresponding reduction in the myoelectric signals (flexion-relaxation phenomenon) from the lumbar muscles. Individuals with pain or injury to the low back are not observed to possess this myoelectric quiescence (Alschuler et al., 2010; Neblett et al., 2010; Shirado et al., 1995), which may indicate injury to or pain within these connective tissues.

The role of the back muscles are divided into two general categories: global and local muscles. Global muscles are usually larger muscles which span multiple regions of the back, such as the erector spinae group (iliocostalis, longissimus, spinalis) and latissimus dorsi. These large muscles enable the trunk to move relative to the global environment through flexion-extension, lateral bending, and twisting. The global muscles also provide tremendous stability to the lumbar region when activated because of their cross sectional area and the number of vertebrae that they span. The second group of muscles is the local muscles, or local stabilizers, and they are much smaller than the global muscles. These muscles (multifidus, rotatores) are believed to provide stability to the neighboring vertebrae and regulate movement between adjacent vertebrae. In conjunction with the back muscles, the abdominal muscles (rectus abdominis, external obliques, internal obliques, and transversus abdominis) aid in the postural support of the trunk, and when co-activated with the parspianl muscles (including quadatus lumborum), provide an anatomical brace to increase spinal stability (Vera-Garcia et al., 2006; 2007). Through the co-activation of the trunk musculature the stiffness of the trunk increases to reduce residual movements that would disrupt static equilibrium of the system. The interrelation of the tissues which compose the trunk profoundly influence how the spine will respond when external perturbations are acting upon the body.

Sensitivity of the Neuromuscular System to Stability Factors of the Spine

The neuromuscular response of the muscles surrounding the spine is dependent upon the margin of stability deemed appropriate by the musculoskeletal system. In other words, the output response of the neural system will conform to the perceived parameters/boundaries established by the structure of the spine. Panjabi et al., (1982) theorizes that mechanical neutral zones exist in the spine where very little or no response from the neuromuscular system is required to sustain the positioning of the vertebrae, relative to one another. Later, Panjabi elaborated on the neutral zone theory in describing the role of stabilization of the spinal system and how three subsystems (active, passive, and neural) were responsible for providing the correct balance in controlling the spinal movements (Panjabi, 1992a; 1992b).

The subsystems are distinct in assisting with spinal stability. The passive tissues (discs, ligaments, fascia) provide a means of support when more static erect postures are performed. The active subsystem involves the muscular control of the spinal movements to either maintain static spinal position, or during dynamic activities. It is also this subsystem that responds to the high amplitude perturbations imposed upon the spine. The passive and active subsystems are also in communication with the nervous system through the afferent receptors embedded within the tissues. The Pacininan corpuscles and Ruffini ending sense tension and length changes within the viscoelastic passive tissues and relay this information to the central nervous system (CNS). Likewise, muscle spindle and bag receptors, which sense length and rate changes in the muscle, as well as Golgi tendon organs, which respond to changes in musculotendonous tension, also provide information to the CNS (Figure 1). All of these sensory inputs are important to sustain spinal stability and act to prevent injury, but which message is more valuable to the body? Since the density of muscle spindle is higher in smaller muscles spanning adjacent vertebral levels it is possible that information from these muscles is better at detecting changes in spinal motion (Boyd-Clark et al., 2002; Brumagne et al., 2000; Nits & Peck, 1986). The mechanoreceptors of the ligaments and fascia are believed to engage in relaying afferent information once finite ends of the range of segment movements are attained, but recent data may suggest that this end range of motion may not be required (Solomonow, 2006; 2011a). Reflexive responses evoked by perturbations to the lumbar spine from both animal and human after viscoelastic creep testing provide data indicating these neuromuscular responses are independent of the information from muscle spindle alone (Olson et al., 2009; Sbriccoli et al., 2007; Solomonow et al., 2003a; 2003b) (Figure 2).

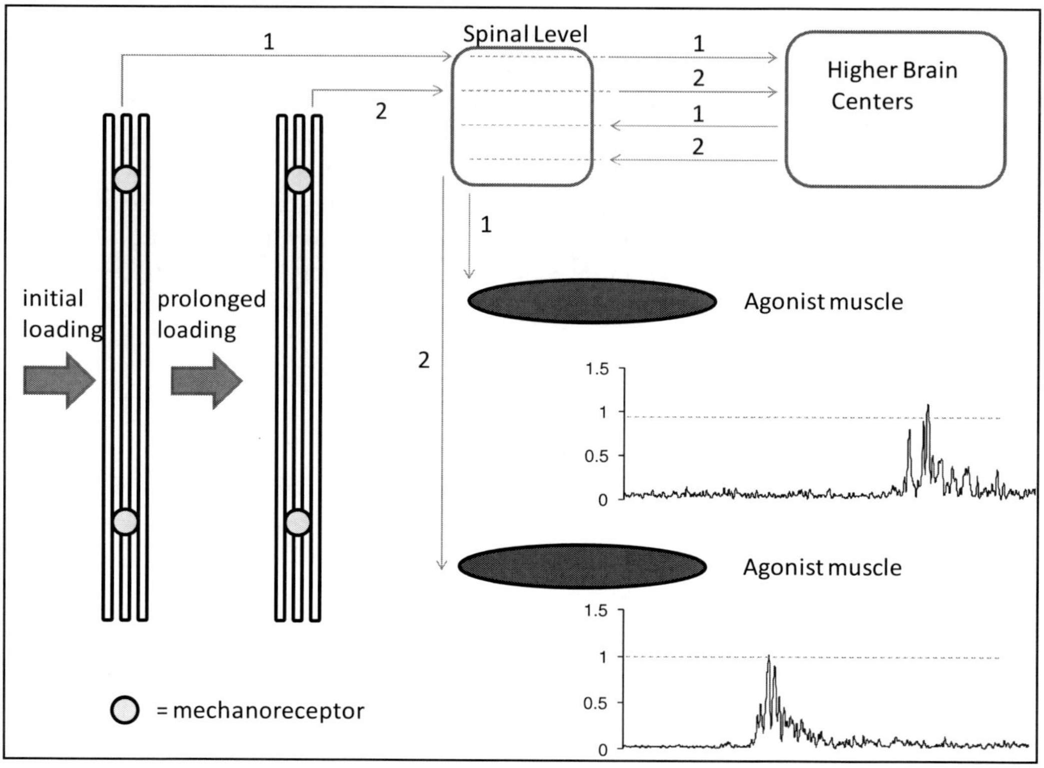

Figure 2. Illustration of response from mechanoreceptors embedded in the ligamentous tissues is initial loading and prolonged loading. The response of the neuromuscular system, in this example amplitude, may be modulated depending upon the mechanical behavior change of the tissues.

The neutral zone itself is very specific to the subsystem being analyzed as neutral zone definitions can be addressed through neuromuscular means, or through purely mechanical conceptualization and calculation. These neutral zones establish a measure of the stability of the system during acute and prolonged loading of the spine. Specifically, the mechanical neutral zone provides mathematically based explanations for the response of the tissues to loading schemes in relation to the boundaries that these tissues (primarily bone) are constrained. For example, the mechanical behaviors of the intervertebral discs are shown to significantly change, but the influences of these changes on the mechanical neutral zones are dependent upon the definition used for the neutral zone (Smit et al., 2011; Wilke et al., 1998). The understanding of the neutral zone forms a basis for evaluating and determining the measures of stability of the spine.

In healthy pain-free individuals, the spinal system maintains a cohesive network of passive control via the intervertebral discs, ligaments, and fascia, active control via the muscles surrounding the joints, and an effective neural (reflexive) response once the system is perturbed. The soft viscoelastic tissues form the passive sub-system used to resist initial perturbations to the spine and allow the confinement of the mechanical neutral zone. As long as these tissues are intact and not mechanically compromised stability of the spine system will be maintained, allowing for slight movements of the vertebrae relative to is adjacent vertebrae. If however, the perturbations imposed onto the system are large and the passive viscoelastic tissues are insufficient to properly resist movement outside the boundaries of the mechanical neutral zone, then the surrounding muscles respond in order to sustain stability of the spine (Cholewicki & McGill, 1996; Panjabi, 2006). Even when small amplitude movements are performed around a neutral sitting posture, the activation patterns of the antagonist to agonist muscles, as ratios, are significantly higher in those with idiopathic chronic low back pain (van Dieën et al., 2003). However, individuals with low back are shown to increase the latency of neuromuscular response when the trunk is perturbed (Radebold et al., 2001; 2000). Cholewicki and colleagues (2005) report increased latency of the paraspinal muscle reflex in collegiate athletes after an injury to the low back occurs, independent of gender or sport. Further, Cholewicki et al (2005) observed a significant increase the recurrence of low back injury (2.8 times, on average) compared to those who had not sustained an injury. Incorporating exercises which focus on training balance, postural control, and loading of the trunk are effective in reducing the reflex latency in those with low back pain (Magnusson et al., 1996; Pedersen et al., 2007; Tsao & Hodges, 2008). Interestingly, when testing proprioceptive impairments to reposition the trunk in those with and without low back pain, no conclusive information can be discerned as differences between populations are highly variable (Gill & Callahan, 1998; O'Sullivan et al., 2003; Newcomer et al., 2000a: 200b, Silfies et al., 2007). However, motion perception appears to be a better indicator of proprioceptive (afferent) impairments compared to repositioning tasks (Lee et al., 2010).

Bipedal Gait and Spinal Loading

As mentioned previously, the evolutionary adaptations that allowed humans to develop terrestrial bipedal locomotion also lead to the increased loading of the spine. Throughout the waking day, when we are mostly aligning out spines with the gravity vector during standing, sitting, and walking activities, the hydrostatic pressure with the intervertebral discs is relatively high resulting in exflux of fluid and reduction of disc height (Laible et al., 1993; Malko et al., 2002; McMillan et al., 1996). The loads on the lumbar spine have been calculated to be in excess of 1.2 times body weight during preferred walking speeds (Cromwell et al., 1989), and at greater walking velocities this loading increases to almost three times body weight (Cheng et al., 1998).

The ground reaction forces (GRFs) acting upon the body during gait directly influence how the body will attenuate and respond to the environment. Many factors influence the GRFs acting on the body during foot contact, such as inertia properties of the body, acceleration of the limb making contact with the support surface, and forward velocity of the body. The body possesses the ability to attenuate these GRFs through joint positioning during landing and cushioning via the heel pad (Whittle, 1999). In fact, lower extremity joints are important in determining the shock attenuation of the GRFs to the pelvis and low back region (Derrick et al., 1998; Edwards et al., 2012; Hamill et al., 2000; McMahon et al., 1985). The neuromuscular system also provides an anticipatory feed-forward mechanism to sustain the trunk and head position in an effort to attenuate the shock of GRFs at heel contact in healthy individuals (Ceccato et al., 2009; Prince et al., 1994). Since knee joint stiffness increases when individuals possess low back pain (Hamill et al., 2009), the neuromuscular system may modify its regulation of temporal recruitment of spinal muscles to effectively attenuate the GRF shock at the spine. Currently, it is unknown whether this modification in motor control strategy is present in those with low back pain and provides an opportunity for further exploration.

Running is generally performed to increase the overall cardiovascular health and promote bone health through loading. The loading of the body may stimulate the osteoblast activity and reduce the potential of osteopenia and osteoarthritis, as well as the production of physiological agents to promote increase vascular function. These loading schemes may have adverse influence on the health of the spine during acute and chronic bouts of exercise. Research into the loading of the spine during running at high, and even moderate intensity, have reported an increase rate of reducing the disc height and volume (Garbutt et al., 1990; Kingsley et al., 2012). Although not correlated specifically with incidents of low back pain, this spinal shrinkage may modify the biomechanical properties of the discs such that the capacity to attenuate shock is reduced. In comparing healthy asymptomatic runners to those with chronic low back pain, Barbutt et al (1990) report similar reductions in spinal stature, however, no specific data with respect to the individual intervertebral discs to illustrate differences between groups are currently present.

Theories of Adaptation to Pain

What influence does pain have on the ability of the neuromuscular system to respond to not only the pain symptom, but in performing tasks of daily living when pain is present? Does the activity of the muscle in proximity to the pain region increase or decrease as a result of noxious stimuli, or do other muscle groups provide compensation to the system? The paradigms provided follow in accord with the two major theories of pain adaptation: the vicious cycle theory and the pain adaptation theory. The vicious cycle theory incorporates further muscle activation to an already painful region in order to reduce movement. As the pain increases, so too does the muscle activity and by doing so the pain is exasperated leading to further complications when attempting to relieve pain symptoms (Roland, 1986). The pain adaptation theory follows a different logic by arguing that muscle activities of muscles near the pain are further inhibited to reduce pain sensation (Lund et al., 1991). Additionally, Hodges and Tucker (2011) suggest that neither of these theories adequately explains the response of the neuromuscular system to nociceptive input from the region of pain. Instead, they postulate a more dynamic theory which takes into consideration a five point evaluation process for pain adaptation: 1) redistribution of muscle activity, 2) mechanical behavior modifications of the tissues, 3) protection from pain, 4) multi-level neuromuscular reorganization, and 5) appraisal of short term and long term modifications to neuromuscular and overall movement performance (Hodges & Tucker, 2011). In further understanding these ideas, empirical evidence from measured muscle activation (Del Santo et al., 2007; Plattner et al., 2011; Plattner et al., 2012) and force generation (Hodges et al., 2008; Salomoni & Graven-Nielsen, 2012) support a more robust theory of pain adaptation. Principally, the maintenance of muscle force during isometric efforts, independent of muscle type, when the rate of motor unit firing decreases is still not fully understood. The pain adaptation theory suggests that agonist muscle activities reduce to compensate for the pain sensations, while also recruiting more motor units from the antagonist muscles to control movement and reduce pain resulting in the reduction of force generation.

The EMG signal detected and recorded as an interpretation of the true biological signal propagating through the muscle can be a useful tool in assessing the adaptation of the muscle to pain. However, caution must be warranted when interpreting these data since the signal is an accumulation of motor unit action potentials, in the case of surface detection. A modified amplitude can be indicative of increased neural drive to the muscle from higher brain centers, but may also be a result of further motor unit recruitment to neighboring or synergist muscles. In addition to the spatial properties of the EMG signal, the frequency content provides a better overall interpretation of the propagating signal. The propagating signal within the muscle is not a direct link to the neural drive relayed via the α-motor neuron, but rather serves to provide information regarding the influence of intrinsic muscle properties. For example, when a muscle is fatigued, usually the force generating capacity of the muscle decreases and the amplitude of the EMG signal increases, but this amplitude change is not always indicative of a fatigued muscle. Further analysis of the power density spectrum of the signal is necessary to determine 1) the power of the signal, and 2) the frequency distribution of the signal. From these data a more complete illustration of the state of neuromuscular fatigue can be established. Similarly, the frequency content of the EMG signal is constructed when muscle is active during episodes of pain to determine how much the muscle is contributing to the force generation. This information, along with analysis of the conduction velocity of the signal via the motor neuron (M-wave, MEP – motor evoked potential) and drive from the higher brain centers (TMR – transcranial magnetic resonance – for the motor areas) is useful in determining how the neuromuscular system responds to pain. Further, the surrounding muscles, synergists as well as antagonist, may contribute to maintaining the force output requirements of the task (Hodges et al., 2008), but with the resulting spatial movement parameters being modified (Mok et al., 2007; Moseley & Hodges, 2006).

Influence of Pain on Muscle Activity and Coordination

The muscle activation patterns of the trunk muscles during locomotion in healthy persons are variable and within milliseconds when observing activation timing within individuals, with inter-individual variability being greater. Nonetheless, there is a distinct neuromuscular pattern followed that is greatly influenced by the external environment in which the individual is navigating and the control from higher brain centers, as well as local control (central pattern generators) (Lacquaniti et al., 1999; Lacquaniti et al., 2002; Prince et al., 1994; Winter, 1983). Changes in the gait velocity, whether determined through perturbations to the system or reorganization of neural drive, result in muscle activation pattern changes (Carlson et al., 1988; Thorstensson et al., 1982; Yang & Winter, 1985). Generally, the amplitude of the EMG signal recorded from the low back muscles increases as the body's velocity increases. In the paraspinal muscles (erector spinae, multifidus) these amplitude changes are readily apparent as the ground reaction forces acting upon the body increase (Anders et al., 2009). Temporal relationships are also influenced by the presence of low back pain as the duration of low back muscle activity is greater when pain is present (Vogt et al., 2003).

The movement coordination between individuals with no back pain (asymptomatic) and those with back pain (symptomatic) is very similar, but there are variation that distinguish the biomechanical parameters between these groups. Spatial and temporal kinematics relationships between asymptomatic and symptomatic groups differ depending upon the severity of the pain and distribution of the pain (Lee et al., 2004; Simmonds et al., 2012). Of course, with shorter stride lengths and reduced velocity in the direction of progression the ground reaction forces acting upon the body are diminished. This may be a coping mechanism used by individuals with low back pain to better attenuate shock during foot contact, which results in reduced movement of the trunk and potentially leading to the alleviation of pain.

Acute Effects

The etiology of low back pain, as mentioned previously, is multifaceted and not fully understood. Since the mechanisms causing the pain are not specifically known, research studying the effects of pain are beneficial in determining how the body adapts to the pain symptoms. In order to provide further information on the effects of back pain during activities of daily living it is important to simulate pain without inducing injury. Common methods of reproducing pain in this manner at the low back are by intramuscular injection of an irritant into (Arendt-Nielsen et al., 1996; Graven-Nielsen et al., 1997; Hodges et al., 2003; Kiesel et al., 2012; Lamoth et al., 2004; Moe-Nilssen et al., 1999; Tsao et al., 2010) or noxious heat application superficially to (Dubois et al., 2011; Tétreau et al., in press) the intended area. This is effective not only because the nociceptors are stimulated, but also because the pain subsides in a relatively short period of time allowing for further analyses of pain recovery. Interestingly, results of injections into spinal ligaments and muscles reveal a long lasting residual pain, even after stretch, when the ligaments are irritated (Tsao et al., 2010). Arendt-Nielsen and colleagues (1996; 1997) provide some of the first insight in regards to the modifications to muscle activation patterns and movement coordination by imposing such procedures during walking gait. Reductions in movement velocity after irritation of the low back tissues is apparent, as well as modified trunk excursion to diminish extraneous motion about the spine in flexion-extension and axial rotation. Similarly, the acute effect of low back pain, in these simulations, leads to reduced acceleration of the trunk which alters the ability of the trunk to attenuate the force transmitted from the support surface to the body (Moe-Nilssen et al., 1999). Muscle activities from the low back are also elevated for ipsilateral foot contacts when experimental pain is induced in healthy individuals (Arendt-Nielsen et al., 1996). Patients with chronic low back pain walk at reduced velocities, have shorter stride length and longer single support durations, but the experimental group in their study did not modify support and swing phase durations. It is possible that the short-duration influence of these injections does not provide sufficient time course to promote a modification in the kinematics control.

The response to movement coordination when low back pain is present requires further analysis of the variables related to the movement. Simple motions during the day are greatly influenced by the perception of pain severity and the body attempts to provide a coping mechanisms to reduce pain in order to perform these daily tasks. Hodges and colleagues (Hodges, 2001; Hodges et al., 2003; Hodges & Richardson, 1997; Hodges & Richardson, 1999; Mok et al., 2007) have extensively researched the role of the trunk musculature when trunk stability is coupled with upper and lower extremity function. The trunk itself must be able to provide stability to the body during many forms of dynamic human movement in order to successfully perform each motor task throughout the day. Moseley and Hodges (2006) indicate that low back pain influences feed-forward mechanisms (namely anticipatory postural adjustments) of control and stability when low back pain is induced in healthy individuals while they perform rapid arm motions. Electromyography from the oblique musculature, in particular, shows invariant onset latencies relative to onset of the deltoid muscles to reduce extraneous movement during the task (Moseley & Hodges, 2006). While the pain in the low back subsides these latencies remain invariant indicating the neuromuscular system remains attuned to reducing potential movement excursion that would harm the body. Similarly, individuals with chronic low back pain are reported to also maintain invariant onset latencies of the oblique muscles during rapid arm movements (Jacobs et al., 2009).

Chronic Effects

Although the acute effects of low back pain can incur drastic changes in the muscle activity and kinematics, the long term modifications with the chronic state of pain bring about an invariant pattern from these same variables. The long term effects of chronic low back pain significantly influence all facets of the movement parameters of locomotion.

Although difference in the muscle activation patterns in spatial and temporal parameters have been illustrated previously between healthy individuals and those with low back pain, the movement synchronizations, by comparison, are also distinct between these groups. Those individuals with low back pain, whether acute or chronic, are shown to perform motions with similar segmental patterning as those without pain. As stated previously, the gait parameters that are significantly different in individuals with chronic low back pain are reduced walking velocity, increased support phase, and reduced swing phase (Lamoth et al,. 2002b; 2006a). The general movement patterns are present in both asymptomatic and symptomatic populations, but as the velocity of the system increases during locomotion there is evidence of divergence of coordinative patterns. The rhythmic movements of the trunk and pelvis are uniquely related during walking and running tasks. At slow speeds, the trunk and pelvis rotate in phase (towards the same direction) in both healthy individuals and those with low back pain (Lamoth et al., 2002a; 2002b; Seay et al., 2011a; Vogt et al., 2001) within the transverse plane. As the velocity of the system increases there is a subtle change in phase relationship between the trunk and pelvis until finally these segments are moving in opposing directions, or antiphase. This antiphase coordination is present at higher walking velocities and also after the body has transitioned into a running gait (Huang et al., 2010; Lamoth et al., 2002a; van Emmerick & Wagenaar, 1996). Individuals with low back pain are shown to not follow this same pattern. In fact, antiphase coordination is not attained in many of these people. Lamoth et al (2002b) report that subjects with low back pain primarily increase rigidity between the trunk and pelvis by maintaining trunk position in space while limiting the transverse rotation of the pelvis as speed increases. Compared to individuals without pain, those with low back pain perform these transverse plane movements with less movement variability, but frontal plane movements are more variant (Lamoth et al., 2006b), indicating less restrictions imposed upon the body by the neuromuscular system. Similarly, van den Hoorn et al (2012) report reduced variability of the trunk-pelvis coordinative pattern compared to asymptomatic persons independent of walking speed (range 0.5 – 1.72 m/s). Seay et al (2011a; 2011b) report differences in the continuous relative phase relationships in the transverse and frontal planes during walking and running between those with low back pain, those with resolved low back pain, and healthy individuals. Overall, the trunk movement between symptomatic and asymptomatic individuals is not different, but the pelvic range of motion about the spinal axis with increased running speed was greater in those with low back pain. This is indicative of a maintained in-phase intersegmental coordination between the trunk and the pelvis during running versus a more anti-phase pattern in asymptomatic individuals. However, during walking the continuous relative phase of the trunk and pelvis are much lower for those with low back pain in comparison to the other groups (Seay et al., 2011b). Greater pelvic range of motion is also documented in those with lumbar disc herniation when attempting to walk faster, and, similar to Seay et al., (2011b), the relative phase between the trunk and pelvis remains more in-phase (Huang et al., 2011). Given the evidence of how individuals with low back pain conform to the environmental constraints imposed upon the body it is unclear what ramifications and further complications this may pose to the other areas of the body.

Further evidence exists to examine how well the system is able to adapt to changes in the environment. Since the body must overcome the constraints imposed by the surrounding environment (sloped walkway, uneven pavement, changing topology, etc.) the capacity for the neuromuscular system and the coordinative patterns of the adjacent segments are required to adapt immediately to these imposed perturbations. Abrupt changes in the gait velocity, controlled in the laboratory, are used to test how the trunk muscles' activation and segmental movement patterns influence the adaptability of the body to these sudden perturbations (Lamoth et al., 2006a).

Discussion/Conclusion

The etiology of low back pain involves multiple factors dependent upon the environmental constraints specific to the task. Variables that are reported to significantly influence the incidence of low back pain include high magnitude loading and high frequency loading, independent of load magnitude. These are issues that are currently being addressed in the literature, but further integration is required with the long term duration of pain symptoms to provide a better understanding of low back health. A clear definitive cause-effect relationship of low back pain cannot be addressed with terrestrial bipedal locomotion since locomotion is not believed to trigger bouts of pain. In fact, bipedal walking in humans is actually reported to be beneficial in the reduction of acute pain symptoms (Malmivaara et al,. 1995). However, the biomechanical and neuromuscular modifications that are present in individuals with chronic low back pain reflect the adaptive capabilities of the body to cope with pain. These neuromuscular and biomechanical adaptations may be acute, along with the pain symptom(s), but with chronic pain affliction the modifications may enhance pain sensation. It is possible to treat low back pain and alleviate pain symptoms though neuromuscular training and coordinative exercises, but it remains unknown how these interventions influence long-term conditions of the biomechanical and motor control variables.

The spine undergoes simultaneous compression, torsion, and bending (flexion-extension, lateral) while performing everyday tasks without incurring significant trauma. Excessive loading involving either increased load magnitude, loading frequency, or the combination is believed to contribute to the biomechanical behavior changes of the passive viscoelastic tissues and vertebrae. Individuals with idiopathic low back pain are able to perform bipedal locomotion without increased pain sensation, but compensation in the movement kinematics although intended to reduce extraneous loading may lead to further complications in the spine and lower extremities. As therapeutic remedies are integrated with motor control training the loading on the spine is modified as the tissues heal. The healing of the tissue would imply a reduction in the pain symptoms leading to modification of the movement parameters closer to those in asymptomatic individuals. However, there is no definitive reversal of the coordinative pattern for kinematics or muscle activation patterns (towards the models of asymptomatic individuals) when pain is alleviated alone. Further research is required to examine the rehabilitation of the low back tissues and reduction of pain in relation to gait parameters. These future research endeavors may also provide additional information regarding the pain adaptation theories currently proposed and lead to support for recent motor control theories of muscle activation and coordination.

All of the aforementioned mechanisms of low back pain, and modifications to biomechanical and motor control parameters, are important to the underlying structural integrity of the spine. As Panjabi (Panjabi 1992a, 1992b; Panjabi et al., 1982) indicates, passive, active, and neurological sub-systems are interconnected to provide stability and sustain the system's state of equilibrium (static or dynamic). When the shock wave from the GRFs acting upon the body during foot contact is transmitted to the spine, the body uses the ligaments, fascia, and bone tissues to attenuate the forces, while the neuromuscular system tunes the muscles to effectively modify muscle tension to provide sufficient joint stiffness. If the mechanical behavior of the ligamentous support surrounding the spine is compromised there appears to be a number of pathways the neuromuscular system has available. The neuromuscular system may increase the neural drive to the primary and secondary muscles, as well as the antagonist muscles, surrounding the area of compromised tissues to prevent excessive spinal movement. This process may be hindered as the reflexive response of the neuromuscular system increases when pain is present in the low back due to the modification of the sensory information input to the CNS from mechanoreceptors embedded within the passive viscoelastic tissues. Since bipedal locomotion, whether walking or running, requires multiaxial loading of the spine and, in turn, multiaxial reaction forces the response of the neuromuscular system may reduce the spine's ability to adequately attenuate GRFs.

In summary, very little is known regarding the long term ramifications of biomechanical and motor control strategies in bipedal locomotion to compensate for low back pain in idiopathic conditions. Data are currently present to suggest modifications to the neuromuscular responses due to irritation, prolonged loading, and potential injury to the passive viscoelastic tissues are evident and require further investigation. It is suggested that future research focus on the recovery of the low back tissues and the alleviation of pain symptoms, and relate these changes to movement and neuromuscular parameter coordination modifications.

References

Adams, M. A., & Dolan, P. (1996). Time dependent changes in the lumbar spine's resistance to bending. *Clinical Biomechanics*, 11, 194-200.

Adams, M. A., & Dolan, P. (2005). Spine biomechanics. *Journal of Biomechanics*, 38, 1972-1983.

Adams, M. A., Mannion, A. F., & Dolan, P. (1999). Personal risk factors for first time low back pain. *Spine*, 24, 2497-2505.

Adams, M. A., Mannion, A. F., & Dolan, P. (1999). Personal risk factors for first time low back pain. Spine, 24, 2497-2505.

Adams, M. A., McNally, D. S., Chinn, H., & Dolan, P. (1994). Posture and the compressive strength of the lumbar spine. *Clinical Biomechanics*, 9, 5-14.

Alschuler, K. N., Neblett, R., Wiggert, E., Haig, A. J., & Geisser, M. E. (2010). Flexion-relaxation and clinical features associated with chronic low back pain: a comparison of different methods of quantifying flexion-relaxation. *Clinical Journal of Pain*, 25, 760-766.

Anders, C., Wagner, H., Puta, C., Grassme, R., & Scholle, H. C. (2009). Healthy humans use sex-specific co-ordination patterns of trunk muscles during gait. *European Journal of Applied Physiology*, 105, 585-594.

Arendt-Nielsen, L., Graven-Nielsen, T., Svarrer, H., & Svensson, P. (1996). The influence of low back pain on muscle activity and coordination during gait: a clinical and experimental study. *Pain,* 64, 231-240.

Bering-Sørenson, F. (1983). A prospective study of low back pain in a general population. *Scandinavian Journal of Rehabilitation and Medicine*, 15, 71-79.

Bogduk, N. Clinical anatomy of the lumber spine and sacrum, third ed., Churchill Livingstone. (1999).

Boyd-Clark, L. C., Briggs, C. A., & Galea, M. P. (2002). Muscle spindle distribution, morphology, and density in longus colli and mutifidus muscles of the cervical spine. *Spine*, 27, 694-701.

Brumagne, S., Cordo, P., Lysens, R., Verschueren, S., & Swinnen, S. (2000). The role of paraspinal muscle spindles in lumbosacral position sense in individuals with and without low back pain. *Spine,* 25, 989-994.

Callahan, J. P., Patla, A. E., & McGill, S. M. (1999). Low back three-dimensional joint forces, kinematics, and kinetics during walking. *Clinical Biomechanics*, 14, 203-216.

Carlson, H., Thorstensson, A., & Nilsson, J. (1988). Lumbar back muscle activity during locomotion: effects of voluntary modifications of normal trunk movement. *Acta Physiologica Scandinavica*, 133, 343-353.

Ceccato, J. C., de Seze, M., Azevedo, C., & Cazalets, J. R. (2009). Comparison of trunk activity during gait initiation and walking in humans. *PLoS ONE* 4(12): e8193. doi:10.1371/journal.pone.0008193

Chazal, J., Tanguy, A., Bourges, M., Gaurel, G., Escande, G., Guillot, M., & Vanneuville, G. (1985). Biomechanical properties of spinal ligaments and a histological study of the supraspinal ligaments in traction. *Journal of Biomechanics*, 18, 167-176.

Cheng, C. K., Chen, H. H., Chen, C. S., & Lee, S. J. (1998). Influences of walking speed changes on the lumbrosacral joint force distribution. *Bio-Medical Materials and Engineering*, 8, 155-165.

Cholewicki, J., & McGill, S. M. (1996). Mechanical stability of the in vivo lumbar spine: implications for injury and chronic low back pain. *Clinical Biomechanics*, 11, 1-15.

Cholewicki, J., Silfies, S. P., Shah, R. A., Greene, H. S., Reeves, N. P., Alvi, K., & Goldberg, B. (2005). Delayed trunk muscle reflex responses increase the risk of low back injuries. *Spine,* 30, 2614-2620.

Cimolin, V., Vismara, L., Galli, M., Zaina, F., Negrini, S., & Capodaglio, P. (2011). Effects of obesity and chronic low back pain on gait. *Journal of Neuroengineering and Rehabilitation*, 8:55 doi:10.1186/1743-0003-8-55

Coenen, P., Kingma, I., Boot, C. R., Bongers, P. M. & van Dieën, J. H. (2012). The contribution of load magnitude and number of load cycles to cumulative low-back load estimates: a study based on in-vitro compression data. *Clinical Biomechanics*, 27, 1083-1086.

Crisco, J. J., & Panjabi, M. (1991). The intersegmental and multisegmental muscles of the lumbar spine: a biomechanical model comparing lateral stabilizing potential. *Spine*, 16, 793-799.

de Looze, M. P., Visser, B., Houting, I., van Rooy, M. A. G,. van Dieën, J. H., & Toussaint, H. M. (1996). Weight and frequency effect on spinal loading in a bricklaying task. *Journal of Biomechanics*, 29, 1425-1433.

Del Santo, F., Gelli, F., Spidalieri, R., & Rossi, A. (2007). Corticospinal drive during painful voluntary contractions at constant force output. *Brain Research*, 1128, 91-98.

Derrick, T. R., Hamill, J., & Caldwell, G. E. (1998). Energy absorption of impacts during running at various stride lengths. *Medicine and Science in Sport and Exercise*, 24, 797-806.

Dolan, P., Mannion, A. F., & Adams, M. A. (1994). Passive tissues help the back muscles to generate extensor moments during lifting. *Journal of Biomechanics*, 27, 1077-1085.

Dubois, J. D., Piché, M., Cantin, V., & Descarreaux, M. (2011). Effect of experimental low back pain on neuromuscular control of the trunk in healthy volunteers and patients with chronic low back pain. *Journal of Electromyography & Kinesiology*, 21, 774-781.

Dumas, G. A., Beaudoin, L., & Drouin, G. (1987). In situ mechanical behavior of posterior spinal ligaments in the lumbar region. An in vitro study. *Journal of Biomechanics*, 20, 301-310.

Edwards, W. B., Derrick, T. R., & Hamill, J. (2012). Musculoskeletal attenuation of impact shock in response to knee angle manipulation. *Journal of Applied Biomechanics*, 28, 502-510.

Fanuele, J. C., Abdu, W. A., Hanscom, B., & Weinstein, J. N. (2002). Association between obesity and functional status in patients with spine disease. *Spine*, 27, 306-312.

Floyd, W. F., & Silver, P. H. S. (1951). Function of erector spinae in flexion of the trunk. *The Lancet,* 260, 133-134.

Floyd, W. F., & Silver, P. H. S. (1951). Function of erector spinae muscles in certain movements in man. *Journal of Physiology*, 129, 184-203.

Frei, H., Oxland, T. R., & Nolte, L. P. (2002). Thoracolumbar spine mechanics contrasted under compression and shear loading. *Journal of Orthopaedic Research*, 20, 1333-1338.

Frymoyer, J. W., Pope, M. H., Costanza, M. C., Rosen, J. C., Goggin, J. E., & Wilder, D. G. (1980). Epidemiological studies of low-back pain, *Spine*, 5, 419-423.

Garbutt, G., Boocock, M. G., Reilly, T, & Troup, J. D. (1990) Running speed and spinal shrinkage in runners with and without low back pain. *Medicine and Science in Sport and Exercise,* 22, 769-772.

Gill, K. P., & Callaghan, M. J. (1998). The measurement of lumbar proprioception in individuals with and without low back pain. *Spine,* 23, 371-377.

Graven-Nielsen, T., P. Svennson, & Arendt-Nielsen, L. (1997). Effects of experimental muscle pain on muscle activity and co-ordination during static and dynamic motor function. *Electroencephalography and clinical Neurophysiology*, 105, 156-164.

Hamill, J., Derrick, T. R., & Holt, K. G. (2000). Shock attenuation and stride frequency during running. *Human Movement Science*, 14, 45-60.

Hamill, J., Moses, M., & Seay, J. (2009). Lower extremity joint stiffness in runners with low back pain. *Research in Sports Medicine*, 17, 260-273.

Hauesler, M., Schiess, R., & Boeni, T. (2011). New vertebral and rib material point to modern bauplan of the Narioktome homo erectus skeleton. *Journal of Human Evolution*, 61, 575-582.

Hindle, R. J., Pearcy, M. J., & Cross, A. (1990). Mechanical function of the human lumbar interspinous and supraspinous ligaments. *Journal of Biomedical Engineering*, 12, 340-344.

Hodges, P. W. (2001). Changes in motor planning of feedforward postural responses of the trunk muscles in low back pain. *Experimental Brain Research*, 141, 261-266.

Hodges, P. W., & Richardson, C. A. (1997). Contraction of the abdominal muscles associated with movement of the lower limbs. *Physical Therapy*, 77, 132-142.

Hodges, P. W., & Richardson, C. A. (1999). Altered trunk muscle recruitment in people with low back pain with upper limb movement at different speeds. *Archive of Physical Medicine and Rehabilitation*, 80, 1005-1012.

Hodges, P. W., & Tucker, K. (2011). Moving differently in pain: a new theory to explain the adaptation to pain. *Pain,* 152, S90-S98.

Hodges, P. W., Ervilha, U. F., & Graven-Nielsen, T. (2008). Changes in motor unit firing rate in synergist muscles cannot explain the maintenance of force during constant force painful contractions. *The Journal of Pain*, 9, 1169-1174).

Hodges, P. W., Moseley, G. L., Gabriellon, A., & Gandevia, S. C. (2003). Experimental muscle pain changes feedforward postural responses of the trunk muscles. *Experimental Brain Research*, 151, 262-271.

Hodges, P.W. (2001). Changes in motor planning of feedforward postural responses of the trunk muscles in low back pain. *Experimental Brain Research,* 141, 261-266.

Huang, Y. P., Bruijn, S. M., Lin, J. H., Meijer, O. G., Wu, W. H., Abbasi-Bafghi, H., Lin, X. C., & van Dieën, J. H. (2011). Gait adaptations in low back pain patients with lumbar disc herniation: trunk coordination and arm swing. *European Spine Journal,* 20, 491-499.

Huang, Y., Meijer, O. G., Lin, J., Bruijn, S. M., Wu, W., Lin, X., Hu, H., Huang, C., Shi, L., & van Dieën, J. H. (2010). The effects of stride length and stride frequency on trunk coordination in human walking. *Gait & Posture*, 31, 444-449.

Indahl, A., Kaigle, A., Reikeras, O., & Holm, S. (1997). Interaction between porcine lumbar intervertebral disc, zygapophysial joints and paraspinal muscles. *Spine*, 22, 2834-2840.

Jacobs, J. V., Henry, S. M., & Nagle, K. J. (2009). People with chronic low back pain exhibit decreased variability in the timing of their anticipatory postural adjustments. *Behavioral Neuroscience,* 123, 455-458.

Jensen, J. N., Holtermann, A., Clausen, T., Mortensen, O. S., Carneiro, I. G., & Andersen, L. L. (2012). The greatest risk for low-back pain among newly educated female health care workers; body weight of physical work load? *BMC Musculoskeletal Disorders*, 13:87, http://www.biomedcentral.com/1471-2474/13/87

Jorgensen, D. J. (2007). Fiscal analysis of emergency admissions for chronic back pain: a pilot study from a Maine hospital. *Pain Medicine*, 8, 354-358.

Kavcic, N., Grenier, S., & McGill, S. M. (2004). Determining the stabilizing role of individual torso muscles during rehabilitation exercises. *Spine,* 29, 1254-1265.

Keller, T. S., Colloca, C. J., Harrison, D. E., Harrison, D. D., & Janik, T. J. (2005). Influence of spine morphology on intervertebral disc loads and stresses in asymptomatic adults: implications for the ideal spine. *Spine Journal*, 5, 297-309.

Kerr, M. S., Frank, J. W., Shannon, H. S., Norman, R. W., Wells, R. P., Neumann, W. P., & Bombardier, C. (2001). Biomechanical and psychological risk factors for low back pain at work. *American Journal of Public Health*, 91, 1069-1075.

Kiesel, K. B., Butler, R. J., Duckworth, A., Halaby, T., Lannan, K., Phifer, C., DeLeal, C., & Underwood, F. B. (2012). Experimentally induced pain alters the EMG activity of the lumbar multifidus in asymptomatic subjects. *Manual Therapy*, 17, 236-240.

Kingsley, M. I., D'Silva, L. A., Jennings, C., Humphries, B., Dalbo, V. J., & Scanlan, A.T. (2012). Moderate-intensity running causes intervertebral disc compression in young adults. *Medicine and Science in Sport and Exercise*, 44, 2199-2204.

Lacquaniti, F., Grasso, R., & Zago, M. (1999). Motor patterns in walking. *News in Physiological Sciences*, 14, 168-174.

Lacquaniti, F., Ivanenko, Y. P., & Zago, M. (2002). Kinematic control of walking. *Archives Italiennes de Biologie*, 140, 263-272.

Laible, J. P., Pfaster, D. S., Krag, M. H., Simon, B. R., & Haugh, L. D. (1993). A poroelastic-swelling finite element model with application to the intervertebral disc. *Spine,* 18, 659-670.

Lamoth, C. J. C., Beek, P. J., Meijer, O. G. (2002a). Pelvis-thorax coordination in the transverse plane during gait. *Gait & Posture*, 16, 101-114.

Lamoth, C. J. C., Daffertshofer, A., Meijer, O. G., & Beek, P. J. (2006a). How do persons with chronic low back pain speed up and slow down? Trunk-pelvis coordination and lumbar erector spinae activity during gait. *Gait & Posture*, 23, 230-239.

Lamoth, C. J. C., Daffertshofer, A., Meijer, O. G., Moseley, G. L., Wuisman, P.I. J.M., & Beek, P. J. (2004). Effects of experimentally induced pain and fear of pain on trunk coordination and back muscle activity during walking. *Clinical Biomechanics*, 19, 551-563.

Lamoth, C. J. C., Daffertshofer, A., Meijer, O.G., Wuisman, P.I.J.M., & Beek, P.J. (2006b). Effects of chronic low back pain on trunk coordination and back muscle activity during walking: changes in motor control. *European Spine Journal*, 15, 23-40.

Lamoth, C. J. C., Meijer, O. G., Wuisman, P.I. J.M., van Dieën, J. H., Levin, M. F., & Beek, P. J. (2002b). Pelvic-thorax coordination in the transverse plane during walking in persons with nonspecific low back pain. *Spine,* 27, E92-E99.

Lee, A. S., Cholewicki, J., Reeves, N. P., Zazulak, B. T., & Mysliwiec, L. W. (2010). Comparison of trunk proprioception between patients with low back pain and healthy controls. *Archives of Physical Medicine and Rehabilitation*, 91, 1327-31.

Lee, C. E., Simmonds, M. J., Etnyre, B. R., & Morris, G. S. (2004). Influence of pain distribution on gait characteristics in patients with low back pain: part 1: vertical ground reaction force. *Spine,* 32, 1329-1336.

Lund, J. P., Donga, R., Widmer, C. G., & Stohler, C. S. (1991). The pain-adaptation model: a discussion of the relationship between chronic musculoskeletal pain and motor activity. *Canadian Journal of Physiology and Pharmacology*, 69, 683-694.

MacKinnon, C. D., & Winter, D. A. Control of whole body balance in the frontal plane during human walking. *Journal of Biomechanics*, 26, 633-644.

Maetzel, A., & Li, L. (2002). The economic burden of low back pain: a review of studies published between 1996 and 2001. *Best Practice & Research Clinical Rheumatology*, 16, 23-30.

Magnusson, M. L., Aleksiev, A., Wilder, D.G., Pope, M.H., Spratt, K., Lee, S.H., Goel, V.K., & Weinstein, J.N. (1996). Unexpected load and asymmetric posture as etiologic factors in low back pain. *European Spine Journal*, 5, 23-35.

Malko, J. A., Hutton, W. C., & Fajman, W. A. (2002). An in vivo MRI study of the changes in volume (and fluid content) of the lumbar intervertebral disc after overnight bed rest and during 8-hour walking protocol. *Journal of Spinal Disorders and Techniques*, 15, 157-163.

Malmivaara, A., Hakkinen, U., Aro, T., Heinrichs, M. L., Koskenniemi, L., Kuosma, E., Lappi, S., Paloheimo, R., Servo, C., Vaaranen, V., Hernberg, S. (1995). The treatment of acute low back pain – bed rest, exercise, or ordinary activity? *New England Journal of Medicine*, 332, 351-355.

Malmqvist, S., Kjaermann, I, Andersen, K., Økland, I., Brønnick, K., & Larsen, J. P. (2012). Prevalence of low back and pelvic pain during pregnancy in a Norwegian population. *Journal of Manipulative and Physiological* Therapy, 35, 272-278.

Maniadakis, N., & Gray, A. (2000). The economic burden of back pain in the UK. *Pain*, 84, 95-103.

Marras, W. S., & Sommerich, C. M. (1991a). A three-dimensional motion model of loads on the lumbar spine. I. Model structure. *Human Factors*, 33, 123-138.

Marras, W. S., & Sommerich, C. M. (1991b). A three-dimensional motion model of loads on the lumbar spine. II. Model validation. *Human Factors*, 33, 139-149.

McGill, S. M., & Brown, S. (1992). Creep response of the lumbar spine to prolonged full flexion. *Clinical Biomechanics*, 7, 43-46.

McGill, S. M., & Kippers, V. (1994). Transfer of loads between lumbar tissues during the flexion-relaxation phenomenon. *Spine*, 19, 2190-2196.

McGill, S. M., & Norman, R. W. (1986). Partitioning the L4-L5 dynamic moment into disc, ligamentous, and muscular components during lifting. *Spine,* 11, 666-678.

McMahon, T. A., Valiant, G., Frederick, E. C. (1985). Groucho running. *Journal of Applied Physiology*, 62, 2326-2337.

McMillan, D. W., Garbutt, G., & Adams, M. A. (1996). Effects of sustained loading on the water content of intervertebral discs: implications for disc metabolism. *Annals of the Rheumatic Diseases*, 55, 880-887.

Millisdotter, M., Stromqvist, B., & Jönsson, B. (2003). Proximal neuromuscular impairment in lumbar disc herniation: a prospective controlled study. *Spine,* 28, 1281-1289.

Moe-Nilssen, R., Ljunggren, A.E., Torebjörk, E. (1999) Dynamic adjustments of walking behavior dependent on noxious input in experimental low back pain. *Pain,* 83, 477-485.

Mok, N. W., Brauer, S. G., & Hodges, P. W. (2007). Failure to use movement in postural strategies leads to increased spinal displacement in low back pain. *Spine,* 32, E537-E543.

Moseley, G. L., & Hodges, P. W. (2006). Reduced variability of postural strategy prevents normalization of motor changes induced by back pain: a risk factor for chronic trouble? *Behavioral Neuroscience*, 120, 474-476.

Neblett, R., Mayer, T. G., Brede, E., Gatchel, R. J. (2010). Correcting abnormal flexion-relaxation in chronic lumbar pain: responsiveness to a new biofeedback training protocol. *Clinical Journal of Pain*, 26, 403-409.

Newcomer, K. L., Laskowski, E. R., Yu, B., Johnson, J. C., & An, K. N. (2000a). Difference in repositioning error among patients with low back pain compared with control subjects. *Spine*, 25, 2488-2493.

Newcomer, K. L., Laskowski, E. R., Yu, B., Larson, D. R., & An, K. N. (2000b). Repositioning error in low back pain: comparing trunk repositioning error in subjects with chronic low back pain and control subjects. *Spine,* 25, 245-250.

Niemitz, C. (2010). The evolution of the upright posture and gait – a review and a new synthesis. *Naturwissenschaften*, 97, 241-263.

Nitz, A. J., & Peck, D. (1986). Comparison of muscle spindle concentrations in large and small human epaxial muscles acting in parallel combinations. *The American Surgeon*, 52, 273-277.

Norman, R., Wells, R., Neumann, P., Frank, J., Shannon, H., & Kerr, M. (1998). Comparison of peak vs. cumulative physical work exposure risk factors for the reporting of low back pain in the automotive industry. *Clinical Biomechanics*, 13, 561-573.

O'Sullivan, P. B., Burnett, A., Floyd, A. N., Gadsdon, K., Logiudice, J., Miller, D., & Quirke, H. (2003). Lumbar repositioning deficit in a specific low back pain population. *Spine,* 28, 1074-1079.

O'Sullivan, P., Phyty, D., Twomey, L., & Allison, G. (1997). Evaluation of specific stabilizing exercise in the treatment of chronic low back pain with radiologic diagnosis of spondylolysis or spondylolisthesis. *Spine*, 22, 2959-2967.

Oleske, D. M., Lavender, S. A., Andersson, G. B. J., Morrissey, M. J., Zold-Kilbourn, P., Allen, C., & Taylor, E. (2006). Risk factors for recurrent episodes of work-related low back disorders in an industrial population. *Spine,* 31, 789-798.

Oleske, D. M., Neelakantan, J., Andersson, G. B., Hinrichs, B. G., Lavender, S. A., Morrissey, M. J., Zold-Kilbourn, P., & Taylor, E. (2004). Factors affecting recovery from work-related, low back disorders in autoworkers. *Archives of Physical Medicine and Rehabilitation*, 85, 1362-1364.

Olson, M. W. (2011). Passive repetitive loading of the lumbar tissues influences force output and EMG during maximal efforts. *European Journal of Applied Physiology*, 111, 1269-1278.

Olson, M. W., Li, L., & Solomonow, M. (2009). Interaction of viscoelastic tissue compliance with lumbar muscles during passive cyclic flexion-extension. *Journal of Electromyography & Kinesiology*, 19, 30-38.

Ostgaard, H. C., Andersson, G. B., & Karlsson, K. (1991). Prevalence of back pain in pregnancy. *Spine,* 16, 549-552.

Panjabi, M. M. (1992). The stabilizing system of the spine. Part I. function, dysfunction, adaptation, and enhancement. *Journal of Spinal Disorders,* 5, 383-389.

Panjabi, M. M. (1992). The stabilizing system of the spine. Part II. neutral zone and instability hypothesis. *Journal of Spinal Disorders*, 5, 390-397.

Panjabi, M. M. (2006). A hypothesis of chronic back pain: ligament subfailure injuries lead to muscle control dysfunction. *European Spine Journal*, 15, 668-676.

Panjabi, M. M., Goel, V. K., & Takata, K. (1982). Physiological strains in the lumbar spinal ligaments. An in vitro biomechanical study. *Spine,* 7, 192-203.

Pedersen, M. T., Essendrop, M., Kotte, J. H., Jorgensen, K., Schibye, B., & Fallentin, N. (2007). Back muscle response to sudden trunk loading can be modified by training among healthcare workers. *Spine*, 32, 1454-1460.

Pintar, F. A., Yoganandan, N., Myers, T., Elhagediab, A., & Sances, A. (1992). Biomechanical properties of the human lumbar spine ligaments. *Journal of Biomechanics,* 25, 1351-1356.

Plattner, K., Baumeister, J., Lamberts, R. P., & Lambert, M. I. (2011). Dissociation in changes in EMG activation during maximal isometric and submaximal low force dynamic contractions after exercise-induced muscle damage. *Journal of Electromyography and Kinesiology*, 21, 542-550.

Plattner, K., Lambert, M. I., Tam, N., & Baumeister, J. (2012). The response of cortical alph activity to pain and neuromuscular changes caused by exercise. *Scandinavian Journal of Medicine & Science in Sport*, doi: 10 111/j.1600-0838.2012.01486.x

Pollintine, P., Dolan, P., Tobias, J. H., & Adams, M. A. (2004a). Intervertebral disc degeneration can lead to "stress shielding" of the anterior vertebral body: a cause of osteoporotic vertebral fracture? *Spine*, 29, 774-782.

Pollintine, P., Przybyla, A. S., Dolan, P., & Adams, M. A. (2004b). Neural arch load-bearing in old and degenerated spines. *Journal of Biomechanics*, 37, 197-204.

Preuschoft, H. (2004). Mechanisms for the acquisition of habitual bipedality: are there biomechanical reasons for the acquisition of upright bipedal posture? *Journal of Anatomy*, 204, 363-384.

Prince, F., Winter, D. A., Stergiou, P., & Walt, S. E. (1994). Anticipatory control of upper body balance during human locomotion. *Gait & Posture*, 2, 19-25.

Radebold, A., Cholewicki, J., Panjabi, M. M., & Patel, T. C. (2000). Muscle response pattern to sudden trunk loading in healthy individuals and in patients with chronic low back pain. *Spine,* 25, 947-954.

Radebold, A., Cholewicki, J., Polzhofer, G. K., & Greene, H. S. (2001). Impaired postural control of the lumbar spine is associated with delayed muscle response times in patients with chronic idiopathic low back pain. *Spine,* 26, 724-730.

Roland, M. (1986). A critical review of the evidence for a pain-spasm-pain cycle in spinal disorders. *Clinical Biomechanics*, 1, 102-109.

Rousseau, M. A., Bradford, D. S., Hadi, T. M., Pedersen, K. L., & Lotz, J. C. (2006). The instant axis of rotation influences facet forces at L5/S1 during flexion/extension and lateral bending. *European Spine Journal*, 15, 299-307.

Salomoni, S. E., & Graven-Nielsen, T. (2012). Experimental muscle pain increases normalized variability of multidirectional forces during isometric contractions. *European Journal of Applied Physiology*, 112, 3607-3617.

Sbriccoli, P., Solomonow, M., Zhou, B., & Lu, Y. (2007). Work to rest duration ratios exceeding unity are a risk factor for low back disorder. *Journal of Electromyography & Kinesiology*, 17, 142-152.

Seay, J. F., Van Emmerick, R. E. A., & Hamill, J. (2011a). Low back pian status affects pelvis-trunk coordination and variability during walking and running. *Clinical Biomechanics*, 26, 572-578.

Seay, J. F., Van Emmerick, R. E. A., & Hamill, J. (2011b). Influence of low back pain status on pelvis-trunk coordination during walking and running. *Spine*, 36, E1070-E1079.

Sharma, M., Langrana, N.A., & Rodriguez, J. (1995). Role of ligaments and facets in lumbar spinal stability. *Spine*, 20, 887-900.

Shirado, O., Ito, T., Kaneda, K., & Strax, T. E. (1995). Flexion-relaxation phenomenon in the back muscles: a comparative study between healthy subjects and patients with chronic low back pain. *American Journal of Physical Medicine & Rehabilitation*, 74, 139-144.

Shirazi-Adl, A., & Panianpour, M. (1999). Effects of changes in lordosis on mechanics of the lumbar spine-lumbar curvature in lifting. *Journal of Spinal Disorders*, 12, 436-447.

Shirazi-Adl, A., & Panianpour, M. (1999). Effects of changes in lordosis on mechanics of the lumbar spine-lumbar velocity and horizontal ground reaction forces in patients with low back pain. *Pain Research and Treatment*, doi:10.1155/2012/214980

Silfies, S. P., Cholewicki, J., Reeves, N. P., & Greene, H. S. (2007). Lumbar position sense and the risk of low back injuries in college athletes: a prospective cohort study. *BMC Musculoskeletal Disorders*, doi:10.1186/1471-2474-8-129

Simmonds, M. J., Lee, C. E., Etnure, B. R., & Morris, G. S. (2012). The influence of pain distribution on walking Adams, M.A., & Dolan, P. (1996). Time dependent changes in the lumbar spine's resistance to bending. *Clinical Biomechanics*, 11, 194-200.

Sjolander, P., Johansson, H., & Djupsjobacka, M. (2002). Spinal and supraspinal effects of activity in ligament afferents. *Journal of Electromyography & Kinesiology*, 12, 167-176.

Smit, T. H., van Tunen, M.S. L.M., van der Veen, A. J., Kingma, I., & van Dieën, J. H. (2011). Quantifying intervertebral disc mechanics: a new definition of neutral zone. *BMC Musculoskeletal Disorders*, 12:38 doi:10.1186/1471-2474-12-38

Solomonow, M. (2006). Sensory-motor control of ligaments and associated neuromuscular disorders. *Journal of Electromyography & Kinesiology*, 16, 549-567.

Solomonow, M. (2011a). Time dependent spine stability: the wise old man and the six blind elephants. *Clinical Biomechanics*, 26, 219-228.

Solomonow, M., Baratta, R. V., Banks, A., Freudenberger, C., & Zhou, B. (2003a). Flexion-relaxation response to static lumbar flexion in males and females. *Clinical Biomechanics*, 18, 273-279.

Solomonow, M., Baratta, R. V., Zhou, B., Burger, E., Zieske, A., & Gendalia, A. (2003b). Muscular dysfunction elicited by creep of lumbar viscoelastic tissues. *Journal of Electromyography & Kinesiology*, 13, 381-396.

Solomonow, M., Zhou, B. H., Baratta, R. V., Lu, Y., & Harris, M. (1999). Biomechanics of increased exposure to lumbar injury caused by cyclic loading: part I. loss of reflexive muscular stabilization. *Spine,* 24, 2426-2434.

Solomonow, M., Zhou, B. H., Lu, Y., & King, K. B. (2011b). Acute repetitive lumbar syndrome: a multi-component insight into the disorder. *Journal of Bodywork & Movement Therapies,* doi:10.1016/j.jbmt.2011.08.005

Solomonow, M., Zhou, B., Baratta, R. V., Zhu, M., & Lu, Y. (2002). Neuromusuclar disorders associated with static lumbar flexion: a feline model. *Journal of Electromyography & Kinesiology,* 12, 81-90.

Svensson, H. O., Vedin, A., Wilhelmsson, C., & Andersson, G. B. J. (1983). Low-back pain in relation to other diseases and cardiovascular factors. *Spine,* 8, 277-285.

Svensson, H.O. & Andersson, G.B.J. (1989). The relationship of low-back pain, work history, work environment, and stress: a retrospective cross-sectional study of 38- to 64- year-old women. *Spine*, 14, 517-522.

Tétreau, C., Dubois, J. D., Piché, M., Descarreaux, M. (in press). Modulation of pain-induced neuromuscular trunk responses by pain expectation: a single group study. *Journal of Manipulative and Physiological Therapeutics,* http://dx.doi.org/10.1016 /j.jmpt.2012.06.008

Thorstensson, A., Carlson, H., Zomlefer, M. R., & Nilsson, J. (1982). Lumbar back muscle activity in relation to trunk movements during locomotion. *Acta Physiologica Scandinavica*, 116, 13-20.

Tsantrizos, A., Ito, K., Aebi, M., & Steffen, T. (2005). Internal strains in healthy and degenerated lumbar intervertebral discs. *Spine*, 30, 2129-2137.

Tsao, H., & Hodges, P. W. (2008). Persistence of improvements in postural strategies following motor control training in people with recurrent low back pain. *Journal of Electromyography & Kinesiology,* 18, 559-567.

Tsao, H., Tucker, K. J., Coppieters, M. W., & Hodges, P. W. (2010). Experimentally induced low back pain from hypertonic saline injections into the interspinous ligament and erector spinae muscle. *Pain*, 150, 167-172.

United States Bureau of Labor Statistics. (2007). Occupational injuries and illnesses: counts, rates, and characteristics. United States Department of Labor, http://www.bls.gov /iif/oshbulletin2006.htm

van den Hoorn, W., Bruijn, S.M., Meijer, O. G., Hodges, P. W., & van Dieën, J. H. (2012). Mechanical coupling between transverse plane pelvis and thorax rotations during gait is higher in people with low back pain. *Journal of Biomechanics*, 45, 342-347.

van Dieën, J. H., Cholewicki, J., & Radebold, A. (2003). Trunk muscle recruitment patterns in patients with low back pain enhances the stability of the lumbar spine. *Spine,* 28, 834-841.

van Dieën, J. H., Weinans, H., & Toussaint, H. M. (1999). Fractures of the lumbar vertebral endplate in the etiology of low back pain: a hypothesis on the causative role of spinal compression in aspecific low back pain. *Medical Hypotheses*, 53, 246-252.

van Emmerick, R. E. A., & Wagenaar, R. C. (1996). Effects of walking velocity on relative phase dynamics in the trunk in humans. *Journal of Biomechanics*, 29, 1175-1184.

Vera-Garcia, F. J., Brown, S. H., Gray, J. R., & McGill, S. M. (2006). Effects of different levels of torso coactivation on trunk muscular and kinematic responses to posteriorly applied sudden loads. *Clinical Biomechanics*, 21, 443-455.

Vera-Garcia, F.J., Elvira, J.L., Brown, S.H., & McGill, S.M. (2007). Effects of abdominal stabilization maneuvers on the control of spine motion and stability against sudden trunk perturbations. *Journal of Electromyography & Kinesiology*, 17, 556-567.

Vogt, L., Pfeifer, K,. Portscher, M., & Banzer, W. (2001). Influences of nonspecific low back pain on three-dimensional lumbar spine kinematics in locomotion. *Spine,* 26, 1910-1919.

Vogt, L., Pfeifer, K., & Banzer, W. (2003). Neuromuscular control of walking with chronic low-back pain. *Manual Therapy*, 8, 21-28.

Voloshin, A. & Wosk, J. (1982). An in vivo study of low back pain and shock absorption in the human locomotor system. *Journal of Biomechanics*, 15, 21-27.

Wang, J. L., Parianpour, M., Shirazi-Adl, A., & Engin, A. E. (2000). Viscoelastic finite-element analysis of a lumbar motion segment in combined compression and sagittal flexion: effect of loading rate. *Spine*, 25, 310-318.

Whittle, M. W. (1999). Generation and attenuation of transient impulsive forces beneath the foot: a review. *Gait & Posture*, 10, 264-275.

Wilke, H. J., Wenger, K., Claes, L. (1998). Testing criteria for spinal implants: recommendations for the standardization of in vitro stability testing of spinal implants. *European Spine Journal*, 7, 148-154.

Wilke, H. J., Wolf, S., Claes, L. E., Arand, M., & Wiesend, A. (1995). Stability increase of the lumbar spine with different muscle groups: a biomechanical in vitro study. *Spine*, 20, 192-198.

Winter, D. A. (1983). Biomechanical motor patterns in normal walking. *Journal of Motor Behavior,* 15, 302-330.

Winter, D. A., MacKinnon, C. D., Ruder, G. K., & Wieman, C. (1993). An integrated EMG/biomechanical model of upper body balance and posture during human gait. *Progress in Brain Research*, 97, 359-367.

Wisleder, D., Smith, M. B., Mosher, T. J., & Zatsiorsky, V. (2001). Lumbar spine mechanical response to axial compression load in vivo. *Spine*, 26, E403-E409.

Woolf, A. D & Pfleger, B. (2003). Burden of major musculoskeletal conditions. *Bulletin of the World Health Organization,* 81, 646-656.

Yahia, H., & Newman, N. (1991). Innervation of spinal ligaments of patients with herniated disc. *Pathology Research Practice*, 187, 936-938.

Yamashita, T., Kanaya, K., Sekine, M., Takebayashi, T., Kawaguchi, S., & Katahira, G. (2002). A quantitative analysis of sensory function in lumbar radiculopathy using current perception threshold testing. *Spine,* 27, 1567-1570.

Yang, A. F., & Winter, D. A. (1985). Surface emg profiles during different walking cadences in humans. *Electroencephalography and clinical Neurophysiology*, 60, 485-491.

In: Gait Biometrics
Editors: Li Li and Matthew Holmes

ISBN: 978-1-62948-381-8

Chapter X

Feedforward and Feedback Control for Gait and Balance

Shuqi Zhang[1,*] ***and Li Li***[2]
[1]Department of Kinesiology, Louisiana State University, Baton Rouge, LA, US
[2]Department of Health and Kinesiology, Georgia Southern University, Statesboro, GA, US

Abstract

Postural control during standing and walking is an important topic when study fall related stability among elderly. Based on the literature and our own data, we suggest the main mechanism that responsible for the determination of postural control during standing among elderly, or patient population, is due to reduced functionality in the sensory system. This includes the sensory organ - tactile, proprioception, vision, vestibular system, sensory nerves conduction velocity and sensory process in the central nervous system. But the main mechanism that responsible for the deterioration of postural control during walking among elderly, or patient population, is due to impairment with in the motor system. Muscle strength, motor nerve function (motor unit recruitment and motor nerve conduction velocity), and coordination among different limbs and the trunk are the main elements that responsible for this mechanism. However, people with peripheral neuropathy have reduced foot sole cutaneous sensation and reduced nerve conduction velocity. This reduction of tactile sensation and slowed signal transmission are highly correlated with the standing balance but to a less extend with walking stability. Rehabilitation effort focused on sensory recovery is more effective than motor rehabilitation among this population. Therefore, postural stability during standing is mainly controlled by a feedback loop, where feedforward control is mainly responsible for postural stability during walking.

* Email: szhan15@tigres.lsu.edu.

Postural Control in Locomotion

Falling, a challenge to balance control during human locomotion is a threat to the quality of life in an aging population. To fully understand the relevant factors of balance control, the neuromuscular systems has to be understood. Normal human locomotion is over ground with periods of one foot contact (walking, running), no feet in contact (running), or both feet in contact (walking, standing) (Winter 1995). Postural control during locomotion requires the real-time interaction with the external environment, because the human body is inherently unstable (Isaacs 1978). Neuromuscular control is required to maintain balance during locomotion. Walking requires dynamic stability control, while standing requires static balance control.

Challenges to postural control may be caused by weak or injured anatomical structures (muscular and skeletal) and neural deficits. Most postural control disorders are characterized by neural deficits. The elderly are particularly prone to many chronic illnesses or disorders that result in rapid degeneration of the nervous system and cause balance disorders. Common underlying chronic diseases are hypertension, diabetes mellitus, arthritis, and heart diseases. In an aging population with balance problems poor postural control may be caused by issues in the peripheral or central nervous systems. For example, in people with Peripheral Neuropathy, balance deficits are mainly due to reduced sensory information from the lower extremity (Simmons, Richardson et al., 1997) and balance issues in Parkinson's Disease are thought to result from central nervous system dysfunction (Ashburn, Stack et al., 2001).

The role of muscle strength in postural control has been investigated among community-dewlling older adults. Muscle strength is considered as an important factor in postural, but the effect of strength training on balance improvement among elderly people is contradicted in different studies. Results of intervention studies, that utilized a strength training intervention, are not consistent as to postural control improvement in elderly with balance disorders. Three bouts of strength training per week for 12 weeks has been shown to improve balance and gait in healthy elderly people (Topp, Mikesky et al., 1993). However, in Buchner's study (1997), short-term strength training did not result in much improvement (Buchner, Cress et al., 1997). In peripheral neuropathy patients, lower extremity strength has not been found to be less than healthy age matched, or correlated with balance issues (Manor and Li 2009)

The nervous system collects and processes information from external and internal environments. After receiving signals from the nervous system, the muscular system generates the necessary amount of force output to maintain balance during movement. The basic neurological mechanism should be understood starting from anatomical structures and their functions in balance control. The functional role of the nervous system may be categorized into three parts: stimulation detection (sensory organs), signal transmission (sensory and motor nerves), and information processing and decision making (CNS). The whole system is responsible for the interaction between the body and the external environment in balance detection and correction.

Visual, vestibular, and somatosensory systems provide sensory information in balance control (Manchester, Woollacott et al., 1989). Sensory activity contributes to movement in two fundamentally different ways. It may mediate error signals following sudden external perturbations and it may influence the pre-programmed motor neural drive in the CNS. Three sensory systems contribute to sensory activity for maintaining balance at all times

(Manchester, Woollacott et al., 1989). However, different sensory systems or different sensation within a sensory system, are weighted differently based on locomotion and task difficulty (Diener and Dichgans 1988; Peterka and Benolken 1995; Simoneau, Ulbrecht et al., 1995; Mergner and Rosemeier 1998). For example, in quiet standing with eyes open and feet apart in a fist width, the somatosensory system and visual system would provide sufficient sensory information to maintain balance as well as to trigger reflexes for balance correction (Timmann, Belting et al., 1994); If the task was conducted on a flexible platform, the vestibular system would be more involved contributing information such as how far from the upright position. The relationship among these three sensory systems is dynamic and changes with sensory information demands and in different postural control circumstances (Lacour and Borel 1993).

In postural control, lower extremity afferent information is mainly from the somatosensory system. The somatosensory system includes the tactile and proprioceptive systems. The tactile system is involved with cutaneous and vibration sensations of the foot sole (J, A et al., 1999). The proprioceptive system includes the muscle spindle and the Golgi tendon organs (GTO). The GTO located within muscles, which monitor the length, velocity, and loading of muscle. This system contributes to the detection of joint position, Joint motion sensation, and kinesthesia (Hijmans, Geertzen et al., 2007; McKeon and Hertel 2007). The roles of these two systems in balance control are not clear, but there is evidence that lower extremity proprioception is not required to trigger balance correction (Bloem, Allum et al., 2002). In the somatosensory system, larger fibers are primarily responsible for muscle spindles and the Golgi tendon organs. However, these are not playing a key role in the control of standing balance compared to smaller fibers which are responsible for innervating the cutaneous receptors (eg. Merkel's cell, Pacinian corpuscles, Meissner's corpuscles, and Ruffini endings in the skin)(Nardone, Grasso et al., 2006; Nardone, Galante et al., 2007). Furthermore, the Meissner's and Pacinian corpuscles (fast-adapting receptors), both responsible for vibrotactile sensitivity are less involved in standing balance control (Nardone, Tarantola et al., 2000; Nardone, Grasso et al., 2006; Shaffer and Harrison 2007). Merkel's cells and Ruffini endings (slow adapting receptors), are responsible for touch and pressure sensitivity and play a key role in standing balance control (Perry, McIlroy et al., 2000). Tactile sensations influence both standing balance, and the control of gait patterns at the ankle, knee, and hip joints as well as the activity of the tibialis anterior muscle during walking (Erickson, Oliver et al., 2004). This further confirms the importance of cutaneous sensation in maintaining balance and the different weights of sensations for various standing tasks.

Somatosensory information is collected by the primary I and II afferents and carried by sensory nerves to spinal cord and cerebellum, and then goes down to the motor unit though motor nerves. The sensory information is processed at lower levels of the spinal cord to trigger corresponding reflexes and motor action or also may go to the upper levels of then spinal cord and cerebellum for more complex processing. This whole loop is called the reflex arc. If the information goes in and out at the same level of spinal cord, it is said to be monosynaptic. Otherwise, it is polysynaptic. Interneurons in the spinal cord may be modulated at higher levels of the CNS, which may modify the neural pathways or reflexes by changing their sensitivity or thresholds in postural control (Clarac, Cattaert et al., 2000). However, the knowledge of central neural mechanisms in postural control is still limited.

The stretch reflex arc contributes to muscle activity in postural control as the connection between the primary I afferent to an α-motoneuron. Feedback triggers different reflexes in corresponding muscles, and in turn, adjusts body posture. Hoffmann reflex could be used to represent the function of stretch reflex arc without the input from muscle spindles (Akazawa, Aldridge et al., 1982; Aldridge and Stein 1982). The latency of the stretch reflex and H-reflex reflect the efficacy of the synaptic transmission between the Ia afferents and motoneurons; while the amplitude of reflex or H-reflex indicates the excitation level of the α-motoneuron. The latency and amplitude of the stretch reflex and H-reflex would be modified by the sensitivity and threshold of the reflex arc, as well as the different central control mechanisms (Capaday and Stein 1986). Although, H-reflex is less affected by the sensory input at the peripheral level, different modulatory effects of the soleus H-reflex is observed in the study of normal people and people with spinal cord injury with different loading stimulation at foot sole (Little and Halar 1985).

Here are examples to illustrate the role for each level of the sensory and central nervous system. When a finger touches hot water, temperature is collected and coded by sensory organs under the skin. The sensory information "hot" is transmitted to the spinal cord and triggers the withdrawal reflex immediately to protect the finger from the bolting water. Also, the sensory information continues to cerebellum to indicate the water was hot. The motor reaction happens immediately after the touch, but before one is cognizant of the hot water. The hand withdrawal is achieved by contraction of the wrist extensor muscles. The regulation of muscle activity is mediated by the stretch reflex. In locomotion, reflexes maintain the gait pattern and conduct balance correction by alternating inhibition or exhibition within different phase of the gait cycle. In gait, soleus muscle reflexes provide inhibition of ankle muscles during the swing phase, and provide excitation in the stance phase. The postural control mechanisms behind the modulation of reflexes will be discussed later.

Fundamental Balance Control in Standing and Gait

Postural control is defined as the act of maintaining, achieving, or restoring a state of balance during any posture or activity (Pollock, Durward et al., 2000). Balance is commonly assessed by examining center of pressure in standing, and dynamic stability is evaluating the variability of the continuous gait pattern in walking. During balance tests, the stance width, stance position, and visual cues influence the results. Balance performance increases with increased stance width (Henry, Fung et al., 2001). However, important to postural control is the different sensory inputs. For example, standing with eyes open results in a better balance performance than with eyes closed (Redfern, Yardley et al., 2001). The elderly have relative poor balance compared to young during standing in any condition (Shumway-Cook, Woollacott et al., 1997). People with peripheral neuropathy, sensory nerve impairments, walk slower and have poor balance. The regulation of postural stability involves both active and passive control in quiet standing and in walking (Winter, Patla et al., 1998; Winter 1995). Passive control refers to locomotion is controlled stiffness, where kinematic proprieties are controlled by anatomical structures as well as gravity. Passive control is considered as a mechanical model built by the inherent human structures. Active control refers to the neural control under real-time interaction with external environment, which requires energy (Bauby and Kuo 2000).

Passive control mainly takes place at the ankle during movement in the anterior-posterior direction and at hip in the mediolateral direction due to the stiffness of the muscles and tendons surrounding the joints (Winter 1995). The passive control of stiffness is explained by the pendulum model in standing as well as the rigid model in walking. The contribution of stiffness would be supported by the decline of postural stability after the lower extremity muscle fatigue. Moreover, it also would help explain consistent postural control under different conditions. However, without active control, passive control is insufficient for stabilization and balance maintenance in standing and walking (Morasso and Schieppati 1999).

Active control is responsible for postural detection and relies on sensory inputs about the interaction between body and external environment. Sensory receptors in visual, vestibular and somatosensory systems are important to postural control (Bauby and Kuo 2000, Qu and Nussbaum 2009). The sensory information helps the CNS make appropriate motor responses. In particular to walking, sensory inputs are important to maintain the dynamic stability in lateral balance (Bauby and Kuo 2000). During quiet standing, sensory input from the lower extremity provides real-time postural detection in all directions.

Motor System

The motor system performs a remarkable analysis each time a movement is made. At the input end there is the command to move the body to achieve some end. In ball-catching for example, the catcher has to simultaneously turn their head and shoulders to the approaching ball, run towards its destination, outstretch an arm and hand, and then wrap the fingers around the ball. At the output end the motor system has to provide a precise time course of contraction and relaxation over several seconds for each of hundreds of muscles. This analysis (in various networks) occurs thorough a series of transformations. The motor system has to break down a command involving the whole body into commands involving small groups of muscle until the drive is specified for each individual muscle.

Active control is achieved by feedback and feedforward control (Seidler, Noll et al., 2004). The motor systems are faced with two conflicting requirements. On the one hand, movement takes place against varying and often unpredictable loads and obstacles. To complete an action, the motor control systems must therefore use sensory feedback to check on the progress and accuracy of the action. On the other hand, movements often have to be made in too short a time to allow for feedback. The motor systems use both feedforward and feedback control for active control in locomotion.

Perhaps a new concept of standing may be dominated by feedback control; while in is dominated by feedforward control. This assertion is supported by popular theories of balance control in standing and gait, the different requirements of sensory information to maintain balance, and adaptations that occur after reduced sensory information from the lower extremities. To sum up, these studies support the hypothesis that standing is dominated by feedback control, which requires more sensory activity to maintain posture; while, in gait, is dominated by feedforward control, which relies less on the sensory activity to maintain the posture.

Theoretical Frameworks of the Standing and Gait

The most popular theories explaining postural control of gait are the Passive Dynamic Theory and Central Pattern Generator Theory. These two theories start from different starting points, but they are in common where the influence of sensory input on balance control during walking is a lesser concern.

Passive Dynamic Theory was developed from a purely mechanical model, in which motion is generated by gravity and inertia. A class of two-legged machines could perform a natural dynamic walking pattern comparable to humans on a shallow slope (McGeer 1990). Once movement is initiated this model is only powered by gravity, which can be combined with active control to work in a more complex environment. According to this theory, the stiffness and the proprieties of human structures explains a majority of control in gait. However, research derived from this theory highlights the importance of active balance control in the medial- lateral direction, especially important for walking (Bauby and Kuo 2000). Two-legged machines with feedforward as the controller and feedback control for error correction was the most stable. Tad's study (1990) provides a detailed mechanical and mathematic analysis of this model. Under the framework of the passive dynamic walking theory, joints stiffness and a series reflex are believed to be foundation of the walking. The stiffness and reflex have a dynamic interacted with external environments by feedforward control at gait generation and feedback control for error correction (Kuo 2002). The Central Pattern Generator Theory, another theory of walking, explains gait control from a neuromuscular perspective. This theory considers gait as a rhythmic movement, which is pre-programmed at the upper level of the CNS (Duysens and Van de Crommert 1998). The required reflexes in gait are pre-set without sensory input as a trigger, evidenced by the decerebrate cat studies (Grillner and Zangger 1979; Pearson and Rossignol 1991). While, sensory input is important in the stance phase, it is less important in the swing phase (Sinkjær, Andersen et al., 2004). The Passive Dynamic and Central Pattern Generator theories have different emphasis. The former starts with a simple machine model and the model is enhanced with feedback and feedforward control. While, the latter one focuses on neural control in walking. No matter which theory, feedback control seems like to have reduced importance in walking.

Unlike gait theories that are less dependent on the influence of sensory activity, standing balance control theories require sensory input. In quiet standing, the center of pressure constantly moves. Hellebrandt (1938) introduced the concept of the stretch reflex strategy, or 'geotropic reflex' where the shift of the Center of Gravity constantly stimulates stretch afferents of postural muscles that then contracted reflexively. This strategy has been questioned since ankle angular motion is unnecessary to elicit a stretch reflex (Kelton & Wright, 1949). Subsequent studies have supported a central organization of posture that does not regulate the ankle angle or muscle length but a more global parameter such as center of gravity position (Gurfinkel et al., 1980; Dietz, 1992). Thus, a systematic strategy has replaced the stretch reflex strategy whose role is limited to coping with perturbations not predicted by the ongoing central program. The goal of maintaining balance is to keep an upright posture, which requires for upright posture and balance error detection and correction in all directions. It is believed that balance error detection relies more on feedback control rather than feedforward control.

Different Requirements for Somatosensory Input in Standing and Gait

Although the goals of postural control in standing and gait are different, the three sensory systems are fundamental to postural control in both conditions. The three systems interact with the external environment and provide information about posture for further analysis in the central nervous system. Somatosensory and vestibular systems are considered to provide more accurate sensory information compared to visual system.

Foot Sole Sensation in Standing and Gait

Somatosensation influences balance control in standing, more than dynamic stability. Plantar pressure directly stimulates the receptors on the sole of the foot. The distribution of plantar pressure is modified by the central nervous system in conjunction with afferent information. Foot sole sensitivity is reported to correlate with plantar pressure distribution (Nurse and Nigg 2001). Regional plantar pressures change after altering the sensory input on foot sole in healthy people in walking and standing (Chen, Nigg et al., 1995; Nurse and Nigg 2001).

People with Peripheral Neuropathy rely on foot sole tactile sensation to provide feedback for maintaining balance and body sway correction (Kars, Hijmans et al., 2009). Poor postural stability is evidenced in people with Peripheral Neuropathy by the exaggerated center of pressure outcomes associated with center of pressure area and the velocity of center of pressure movement (Kars, Hijmans et al., 2009). However, the role of foot cutaneous sensation to maintain balance in healthy people is moderate (Meyer, Oddsson et al., 2004). The different perspectives on the role of foot cutaneous sensation in balance control may be explained by a declined sensory system in people with Peripheral Neuropathy. Foot sole tactile sensation influences plantar pressure distribution in standing suggesting that cutaneous input plays an important role in standing balance control for people with Peripheral Neuropathy. Feedback from foot sole sensation, among other sensory information, is vitally important for balance control among this population.

Many studies have investigated the effect of acutely reduced sensation on the plantar pressure distribution in healthy people by inducing insensitivity on the whole, fore, or rear foot. Reduced vibration sensitivity threshold at the hallux has been linked to reduced pressure in the same region among healthy people in walking and running (Chen, Nigg et al., 1995; Nurse and Nigg 2001; Caselli, Pham et al., 2002; Eils, Nolte et al., 2002; Eils, Behrens et al., 2004). It has been shown that reduced sensation in the whole foot surface leads to decreased peak pressure at the big toe and heel, and another has shown that the center of pressure shifts the load to central forefoot, lateral forefoot and lateral midfoot in walking (Eils, Nolte et al., 2002; Eils, Behrens et al., 2004). However, recent results, using a local anesthetic solution, has put elicited a new concept that plantar pressure distribution in gait is not affected by targeted reduced cutaneous sensation (Höhne, Stark et al., 2009). These inconsistent findings may be explained by different distributions of reduced sensation areas (Höhne, Stark et al., 2009; Kars, Hijmans et al., 2009).

Perhaps the effect of reduced foot sole sensation on plantar pressure distribution is specific to standing, but not to walking (Zhang and Li 2012). Hohne's study, which targeted plantar sensory afferents, with effectively reduced sensitivity, among healthy people using intradermal injections of anesthetic solution, but plantar pressure distribution in walking did not significantly change. Peripheral Neuropathy induced foot sole sensation loss did not affect plantar pressure distribution in walking (See figure 1 for more detail). This may be because walking is less dependent on feedback but more on feedforward control, just like in the central pattern generator theory (Rossignol, Dubuc et al., 2006).

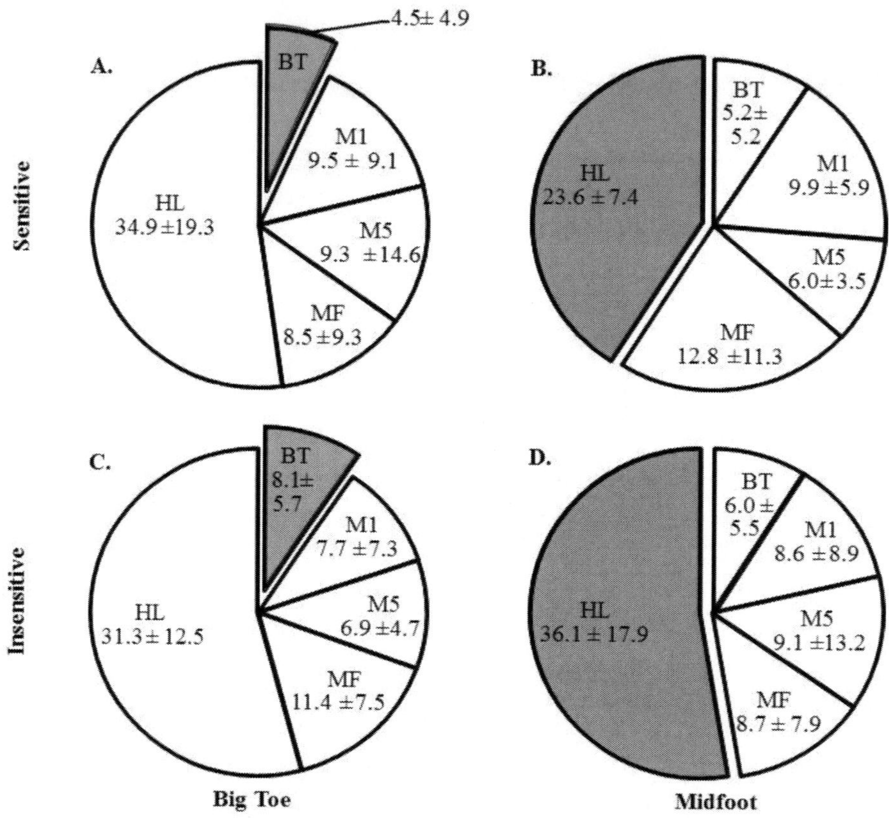

Figure 1. The sensitivity of big toe (BT) and Midfoot (MF) effects on plantar pressure distribution significantly during standing. The relative peak pressure of the five tested sites (BT, MF, M1 – first metatarsal, M5 – fifth metatarsal, and HL – Heel are presented here for four different situations, respectively. Left column shows BT sensitivity impairment led to different overall plantar pressure distributions ($P < .05$), where insensitive BT caused greater pressure under BT ($P < .05$). Right column presents the differential effects of MF sensitivity ($P < .05$) on overall plantar pressure distributions where insensitive MF caused greater pressure under HL ($P < .05$). The gray area in charts A & C (B & D) signify the significant differences.

Foot sole sensation affects plantar pressure distribution in standing but not in walking, which indicates cutaneous input plays less of a role in walking among people with Peripheral Neuropathy. Studies of spinal and decerebrate cats suggest that quadruped locomotion could occur without feedback control (Armstrong 1986). Rhythmic movements like walking could be performed in the absence of descending or afferent inputs (Grillner and Zangger 1979;

Pearson and Rossignol 1991); while bipedal gait is generally presumed to require some level of cerebral control (Reisman, Wityk et al., 2007). In some studies, minimal movement was observed to have been evoked by motor responses caused by sudden interference from the external environment (Lam, Anderschitz et al., 2006), which implied that feedback control was not necessary to maintain gait but only helps when the external environment becomes challenged. These observations suggest that peripheral feedback (foot cutaneous sensory information) doesn't influence foot pressure distribution and walking is mainly modulated by feedforward control rather than feedback control.

Different Modulations of H-Reflex in Standing and Walking

The regulation of motoneuron activity is a result of supraspinal as well as spinal inputs and it is of interest to investigate the excitability at the peripheral level, namely in the spinal reflex pathway. It is accepted that a submaximal stimulation produces a characteristic reflex response (H wave), which is the result of the motoneuron discharge evoked by the activation of the Ia fibers from the muscle spindles. Increasing the stimulus intensity results in the direct depolarization of the α-motoneurons' axons (Maffiuletti, Martin et al., 2001). The ratio between the amplitude of the maximum reflex response (H_{max}) and maximum direct motor wave (M_{max}) is commonly used as an index for estimating the level of reflex excitability of the motor pool (H_{max}-to M_{max} ratio). This depends on the facilitation of the transmission between the Ia fibers and the α-motoneurons (Schieppati 1987). An H-reflex test provides a better measure of the stretch reflex, and it reflects any changes in synaptic efficacy between the muscle afferents and the α-motoneurons. Its function is modulated by the central mechanism and the modulation is accompanied by changes in both reflex sensitivity and reflex threshold.

Capaday and Stein's study shows walking requires more modulation of the soleus H-reflex compared to standing. Greater reflexes observed during standing is consistent with the control of position required to maintain a stable posture. However, the amplitude of the H-reflex changes with different phases in walking, with small reflexes during the swing phase and large reflexes during the stance phase. This indicates that there are different modulation requirements in standing and walking, suggesting different neural control mechanisms. Walking requires a certain amount of compliance (Houk 1976), whereas standing may require a more rigid control of ankle position. The consistently large reflexes in standing imply the importance of feedback control to the standing phase during walking. In walking, the small reflex amplitudes are partly due to the required activation of the soleus for dorsiflexion. The spontaneous reflex increases and decreases between stance and swing phases, as well as the earlier onset of muscle activity than latency of the H-reflex, suggests that this process is pre-programmed by the CNS and may be subject to feedforward control rather than feedback control.

In general, the conduction velocity of the reflex arc is assessed with the sensory nerve conduction velocity and motor nerve conduction velocity separately, rather than being studied as a function of the entire loop. The impaired motor nerve conduction velocity is correlated with the force output, whereas impaired sensory nerve conduction velocity is correlated with the body sway area in quiet standing and walking speed in gait (Nardone, Tarantola et al., 2000).

Details of Short and Long Latency (Large Ia and Small II Afferents) in Soleus Muscle

The role of primary Ia and secondary II afferents has been investigated in standing and walking. In the seated position, the stretch of the soleus muscle leads to two bursts. A short latency response with an onset of approximately 40 ms that is attributed to monosynaptic excitation of spinal motoneurons from the large diameter group Ia afferent fibers (Taylor, Stein et al., 1985; Matthews 1991). Along latency response of approximately 70 ms that is mediated by group II afferent fibers (Schieppati and Nardone 1997). In standing, both types of afferents contribute to balance control in healthy people. The comparable importance of these two types of afferents is only observed during stance phase in walking. Moreover, ischemia did not affect primary Ia afferents function in walking, which suggests that overall, group II afferents contribute more in walking (Grey, Ladouceur et al., 2001).

Adaptation Strategy after Reduced Somatosensation in Standing and Walking

When sensory input is reduced, the nervous system compensates to maintain normal posture in locomotion. The compensation strategies occur at both the peripheral and central nervous systems. The compensation in the peripheral sensory system is achieved by reweighting contribution of the multiple sensation system within the feedback control, which is more applicable to the balance control in standing. While, the central mechanism adaptation would be achieved by new central mechanism in feedforward control, which is more applied to the dynamic stability in walking.

Walking

One approach to understanding locomotor adaptation is to examine motor after effects because the presence of after effects indicates that adaptation has indeed taken place (Lam, Anderschitz et al., 2006). In a young healthy population, experimentally reduced plantar sensation doesn't significantly influence dynamic stability during walking or even during perturbed walking (Dingwell and Cavanagh 2001; Höhne, Stark et al., 2011). For people with peripheral neuropathy, reduced somatosensory input would decrease the walking speed and postural stability (Manor, Wolenski et al., 2008). Current study about the adaptation in gait and its effects in this population indicates this population adopts different feed-forward gait compensatory strategies in response to peripheral sensory loss (Bunday and Bronstein 2009).

Standing

It is well known that intact subjects actively seek out alternative sources of sensory information in destabilizing environments while patients compensate for sensory deficits to stabilize posture (Nashner 1982). The process of sensory re-weighting (inter-modality) is so termed because of strong coupling of one stimulus with postural sway (up-weighting) in conjunction with weak coupling with another stimulus (down-weighting) (Oie, Kiemel et al., 2002; Peterka 2002). Reduced somatosensory will affect the behaviors of the center of pressure in young and elderly people. Plantar stimulations improved postural control in the balance test with standing on a firm surface and in absence of vision (Maki, Perry et al., 1999; Maurer, Mergner et al., 2001; Priplata, Niemi et al., 2003; Nurse, Hulliger et al., 2005; Vaillant, Vuillerme et al., 2008). The balance control would be enhanced by the increasing sensitivity of vestibular and vibrated sensation in young people (Horak and Hlavacka 2001; Preszner-Domjan, Nagy et al., 2012). In healthy young people, increased vibrating sensation would compensated to the reduced sensory input at foot sole in the eyes closed condition (Preszner-Domjan, Nagy et al., 2012). The strategy of multisensory reweighting occurs within the somatosensory system overcomes the lack of sensory input. In people with peripheral neuropathy, the reweighting sensation within somatosensory system is not enough to provide sufficient sensory information for balance control; it requires involving the other two sensory systems. Reweighing multiple sensation system is an important strategy of balance control in standing. It would reweight within somatosensory system or with other two systems.

With pathologies that affect balance control in postural and gait, adaptive strategy in gait is mainly in the forms of altered feedforward control; while multisensory reweighting is dominant adaptive strategy in standing. They support the hypothesis that gait relies more on feedforward control, but standing relies more on feedback control.

References

Akazawa, K., J. Aldridge, et al., (1982). "Modulation of stretch reflexes during locomotion in the mesencephalic cat." *J. Physiol.* 329(1): 553-567.

Aldridge, J. and R. Stein (1982). "Nonlinear properties of stretch reflex studied in the decerebrate cat." *Journal of neurophysiology* 47(2): 179-192.

Armstrong, D. M. (1986). "Supraspinal contributions to the initiation and control of locomotion in the cat." *Prog. Neurobiol.* 26(4): 273-361.

Ashburn, A., E. Stack, et al., (2001). "A community-dwelling sample of people with Parkinson's disease: characteristics of fallers and non-fallers." *Age and Ageing* 30(1): 47-52.

Bauby, C. E. and A. D. Kuo (2000). "Active control of lateral balance in human walking." *Journal of biomechanics* 33(11): 1433-1440.

Bloem, B. R., J. H. Allum, et al., (2002). "Triggering of balance corrections and compensatory strategies in a patient with total leg proprioceptive loss." *Exp. Brain Res.* 142(1): 91-107.

Buchner, D. M., M. E. Cress, et al., (1997). "The effect of strength and endurance training on gait, balance, fall risk, and health services use in community-living older adults." *The*

Journals of Gerontology Series A: Biological Sciences and Medical Sciences 52(4): M218.

Bunday, K. L. and A. M. Bronstein (2009). "Locomotor adaptation and aftereffects in patients with reduced somatosensory input due to peripheral neuropathy." *J. Neurophysiol.* 102(6): 3119-3128.

Capaday, C. and R. Stein (1986). "Amplitude modulation of the soleus H-reflex in the human during walking and standing." *The Journal of neuroscience* 6(5): 1308-1313.

Caselli, A., H. Pham, et al., (2002). "The forefoot-to-rearfoot plantar pressure ratio is increased in severe diabetic neuropathy and can predict foot ulceration." *Diabetes Care* 25(6): 1066-1071.

Chen, H., B. M. Nigg, et al., (1995). "Influence of sensory input on plantar pressure distribution." *Clin. Biomech. (Bristol, Avon)* 10(5): 271-274.

Clarac, F., D. Cattaert, et al., (2000). "Central control components of a 'simple'stretch reflex." Trends in neurosciences 23(5): 199-208.

Diener, H. and J. Dichgans (1988). "On the role of vestibular, visual and somatosensory information for dynamic postural control in humans." Prog Brain Res 76: 253-262.

Dingwell, J. B. and P. R. Cavanagh (2001). "Increased variability of continuous overground walking in neuropathic patients is only indirectly related to sensory loss." *Gait & Posture* 14(1): 1-10.

Duysens, J. and H. W. Van de Crommert (1998). "Neural control of locomotion; Part 1: The central pattern generator from cats to humans." Gait & Posture 7(2): 131-141.

Eils, E., S. Behrens, et al., (2004). "Reduced plantar sensation causes a cautious walking pattern." *Gait Posture* 20(1): 54-60.

Eils, E., S. Nolte, et al., (2002). "Modified pressure distribution patterns in walking following reduction of plantar sensation." *J. Biomech.* 35(10): 1307-1313.

Erickson, M. A., T. Oliver, et al., (2004). "Biomechanical assessment of conventional unit rod fixation versus a unit rod pedicle screw construct: a human cadaver study." *Spine (Phila Pa 1976)* 29(12): 1314-1319.

Grey, M. J., M. Ladouceur, et al., (2001). "Group II muscle afferents probably contribute to the medium latency soleus stretch reflex during walking in humans." *J. Physiol.* 534(3): 925-933.

Grillner, S. and P. Zangger (1979). "On the central generation of locomotion in the low spinal cat." *Experimental brain research* 34(2): 241-261.

Henry, S. M., J. Fung, et al., (2001). "Effect of stance width on multidirectional postural responses." *J. Neurophysiol.* 85(2): 559-570.

Hijmans, J. M., J. H. Geertzen, et al., (2007). "A systematic review of the effects of shoes and other ankle or foot appliances on balance in older people and people with peripheral nervous system disorders." *Gait Posture* 25(2): 316-323.

Höhne, A., C. Stark, et al., (2009). "Plantar pressure distribution in gait is not affected by targeted reduced plantar cutaneous sensation." *Clinical biomechanics (Bristol, Avon)* 24(3): 308.

Höhne, A., C. Stark, et al., (2011). "Effects of reduced plantar cutaneous afferent feedback on locomotor adjustments in dynamic stability during perturbed walking." *Journal of biomechanics* 44(12): 2194-2200.

Horak, F. B. and F. Hlavacka (2001). "Somatosensory loss increases vestibulospinal sensitivity." *J. Neurophysiol.* 86(2): 575-585.

Houk, J. C. (1976). "An Assessment of Stretch Reflex Function." *Understanding the Stretch* Reflex 44: 303.

Isaacs, B. (1978). "Are falls a manifestation of brain failure?" Age and ageing 7(suppl): 97-111.

J, B. J., C. P. A, et al., (1999). Human Physiology, Blackwell Science Ltd, Osney Mead, Oxford

Kars, H. J., J. M. Hijmans, et al., (2009). "The effect of reduced somatosensation on standing balance: a systematic review." *J. Diabetes Sci. Technol.* 3(4): 931-943.

Kuo, A. D. (2002). "The relative roles of feedforward and feedback in the control of rhythmic movements." Motor control-champing 6(2): 129-145.

Lacour, M. and L. Borel (1993). "Vestibular control of posture and gait." Archives italiennes de biologie 131(2): 81-104.

Lam, T., M. Anderschitz, et al., (2006). "Contribution of feedback and feedforward strategies to locomotor adaptations." *J. Neurophysiol.* 95(2): 766-773.

Little, J. and E. Halar (1985). "H-reflex changes following spinal cord injury." *Archives of physical medicine and rehabilitation* 66(1): 19.

Maffiuletti, N. A., A. Martin, et al., (2001). "Electrical and mechanical H(max)-to-M(max) ratio in power- and endurance-trained athletes." *J. Appl. Physiol.* 90(1): 3-9.

Maki, B. E., S. D. Perry, et al., (1999). "Effect of facilitation of sensation from plantar foot-surface boundaries on postural stabilization in young and older adults." *The Journals of Gerontology Series A: Biological Sciences and Medical Sciences* 54(6): M281-M287.

Manchester, D., M. Woollacott, et al., (1989). "Visual, vestibular and somatosensory contributions to balance control in the older adult." Journal of Gerontology 44(4): M118-M127.

Manor, B., P. Wolenski, et al., (2008). "Faster walking speeds increase local instability among people with peripheral neuropathy." *Journal of biomechanics* 41(13): 2787-2792.

Manor, B. and L. Li (2009). "Characteristics of functional gait among people with and without peripheral neuropathy." Gait & posture 30(2): 253-256.

Matthews, P. B. (1991). "The human stretch reflex and the motor cortex." Trends Neurosci 14(3): 87-91.

Maurer, C., T. Mergner, et al., (2001). "Human balance control during cutaneous stimulation of the plantar soles." *Neuroscience letters* 302(1): 45-48.

McGeer, T. (1990). "Passive dynamic walking." The International Journal of Robotics Research 9(2): 62-82.

McKeon, P. O. and J. Hertel (2007). "Diminished plantar cutaneous sensation and postural control." *Percept Mot. Skills* 104(1): 56-66.

Mergner, T. and T. Rosemeier (1998). "Interaction of vestibular, somatosensory and visual signals for postural control and motion perception under terrestrial and microgravity conditions—a conceptual model." Brain Research Reviews 28(1): 118-135.

Meyer, P. F., L. I. Oddsson, et al., (2004). "The role of plantar cutaneous sensation in unperturbed stance." *Exp. Brain Res.* 156(4): 505-512.

Morasso, P. G. and M. Schieppati (1999). "Can muscle stiffness alone stabilize upright standing?" *J. Neurophysiol.* 82(3): 1622-1626.

Nardone, A., M. Galante, et al., (2007). "Balance control in Sensory Neuron Disease." *Clin. Neurophysiol.* 118(3): 538-550.

Nardone, A., M. Grasso, et al., (2006). "Balance control in peripheral neuropathy: are patients equally unstable under static and dynamic conditions?" *Gait Posture* 23(3): 364-373.

Nardone, A., J. Tarantola, et al., (2000). "Loss of large-diameter spindle afferent fibres is not detrimental to the control of body sway during upright stance: evidence from neuropathy." *Exp. Brain Res.* 135(2): 155-162.

Nardone, A., J. Tarantola, et al., (2000). "Loss of large-diameter spindle afferent fibres is not detrimental to the control of body sway during upright stance: evidence from neuropathy." *Experimental brain research* 135(2): 155-162.

Nashner, L. M. (1982). " Analysis of Stance Posture in Humans." *Handbook of Behavioral Neurobiology (New York: Plenum Press.*

Nurse, M. A., M. Hulliger, et al., (2005). "Changing the texture of footwear can alter gait patterns." *J. Electromyogr. Kinesiol.* 15(5): 496-506.

Nurse, M. A. and B. M. Nigg (2001). "The effect of changes in foot sensation on plantar pressure and muscle activity." *Clin. Biomech. (Bristol, Avon)* 16(9): 719-727.

Oie, K. S., T. Kiemel, et al., (2002). "Multisensory fusion: simultaneous re-weighting of vision and touch for the control of human posture." *Cognitive Brain Research* 14(1): 164-176.

Pearson, K. G. and S. Rossignol (1991). "Fictive motor patterns in chronic spinal cats." *J. Neurophysiol.* 66(6): 1874-1887.

Perry, S. D., W. E. McIlroy, et al., (2000). "The role of plantar cutaneous mechanoreceptors in the control of compensatory stepping reactions evoked by unpredictable, multi-directional perturbation." *Brain Res.* 877(2): 401-406.

Peterka, R. J. and M. S. Benolken (1995). "Role of somatosensory and vestibular cues in attenuating visually induced human postural sway." Experimental Brain Research 105(1): 101-110.

Peterka, R. (2002). "Sensorimotor integration in human postural control." *Journal of neurophysiology* 88(3): 1097-1118.

Pollock, A. S., B. R. Durward, et al., (2000). "What is balance?" Clinical rehabilitation 14(4): 402-406.

Preszner-Domjan, A., E. Nagy, et al., (2012). "When does mechanical plantar stimulation promote sensory re-weighing: standing on a firm or compliant surface?" *Eur. J. Appl. Physiol.* 112(8): 2979-2987.

Preszner-Domjan, A., E. Nagy, et al., (2012). "When does mechanical plantar stimulation promote sensory re-weighing: standing on a firm or compliant surface?" *European journal of applied physiology*: 1-9.

Priplata, A. A., J. B. Niemi, et al., (2003). "Vibrating insoles and balance control in elderly people." *Lancet* 362(9390): 1123-1124.

Qu, X. and M. A. Nussbaum (2009). "Evaluation of the roles of passive and active control of balance using a balance control model." Journal of biomechanics 42(12): 1850-1855.

Redfern, M. S., L. Yardley, et al., (2001). "Visual influences on balance." *Journal of anxiety disorders* 15(1): 81-94.

Reisman, D. S., R. Wityk, et al., (2007). "Locomotor adaptation on a split-belt treadmill can improve walking symmetry post-stroke." *Brain* 130(Pt 7): 1861-1872.

Rossignol, S., R. Dubuc, et al., (2006). "Dynamic sensorimotor interactions in locomotion." *Physiol. Rev.* 86(1): 89-154.

Schieppati, M. (1987). "The Hoffmann reflex: a means of assessing spinal reflex excitability and its descending control in man." *Prog. Neurobiol.* 28(4): 345-376.

Schieppati, M. and A. Nardone (1997). "Medium-latency stretch reflexes of foot and leg muscles analysed by cooling the lower limb in standing humans." *J. Physiol.* 503 (Pt 3): 691-698.

Seidler, R., D. Noll and G. Thiers (2004). "Feedforward and feedback processes in motor control." Neuroimage 22(4): 1775-1783.

Shaffer, S. W. and A. L. Harrison (2007). "Aging of the somatosensory system: a translational perspective." *Phys. Ther.* 87(2): 193-207.

Shumway-Cook, A., M. Woollacott, et al., (1997). "The effects of two types of cognitive tasks on postural stability in older adults with and without a history of falls." *J. Gerontol. A Biol. Sci. Med. Sci.* 52(4): M232-240.

Simmons, R. W., C. Richardson, et al., (1997). "Postural stability of diabetic patients with and without cutaneous sensory deficit in the foot." *Diabetes research and clinical practice* 36(3): 153-160.

Simoneau, G. G., J. S. Ulbrecht, et al., (1995). "Role of somatosensory input in the control of human posture." Gait & posture 3(3): 115-122.

Sinkjær, T., J. B. Andersen, M. Ladouceur, L. O. Christensen and J. B. Nielsen (2004). "Major role for sensory feedback in soleus EMG activity in the stance phase of walking in man." J Physiol 523(3): 817-827.

Taylor, J., R. Stein, et al., (1985). "Impulse rates and sensitivity to stretch of soleus muscle spindle afferent fibers during locomotion in premammillary cats." *Journal of neurophysiology* 53(2): 341-360.

Timmann, D., C. Belting, et al., (1994). "Influence of visual and somatosensory input on leg EMG responses in dynamic posturography in normals." Electroencephalography and Clinical Neurophysiology/Evoked Potentials Section 93(1): 7-14.

Topp, R., A. Mikesky, et al., (1993). "The effect of a 12-week dynamic resistance strength training program on gait velocity and balance of older adults." *The Gerontologist* 33(4): 501-506.

Vaillant, J., N. Vuillerme, et al., (2008). "Effect of manipulation of the feet and ankles on postural control in elderly adults." *Brain Res. Bull* 75(1): 18-22.

Winter, D. (1995). "Human balance and posture control during standing and walking." *Gait & Posture* 3(4): 193-214.

Winter, D. A., A. E. Patla, et al., (1998). "Stiffness control of balance in quiet standing." *J. Neurophysiol.* 80(3): 1211-1221.

Zhang, S. and L. Li (2012). "The differential effects of foot sole sensory on plantar pressure distribution between balance and gait." *Gait & Posture*. 37(4): 532-5.

In: Gait Biometrics
Editors: Li Li and Matthew Holmes
ISBN: 978-1-62948-381-8

Chapter XI

Biometric Encryption for Gait Recognition

A. M. Hamel*, J. M. R. Berrigan, C. Klempner, J. Rulloda, B. Vollbrecht, S. Yee and S. D. Perry
Department of Physics and Computer Science,
Department of Kinesiology and Physical Education,
Wilfrid Laurier University, Waterloo, Ontario, Canada

Abstract

Biometrics are revered as strong, secure, and practically unforgeable forms of authentication. However, one aspect that is often overlooked is privacy. Although a biometric may be difficult to forge or steal, the image of it stored in a database may be vulnerable to viewing or reuse by unauthorized persons, or to theft for the purposes of a replay attack. One solution to the privacy issue is the use of biometric encryption or biometric templates. This technique, proposed in 1994 by Tomko et al., stores only a template of the original biometric, instead of a verbatim digital copy of it. This has the additional advantage that, if the database is compromised, the existing template can be revoked and a new template created from the original biometric. A number of biometrics--such as iris, voice, fingerprint, and face--have been used with biometric encryption, and there have been some efforts to use gait in this context, particularly with mobile devices. In this paper we discuss the techniques and implementations of biometric encryption, survey the existing work utilizing gait data with biometric encryption, and propose some future work in using gait data with this technique.

* E-mail address: ahamel@wlu.ca.

Introduction

Biometrics are revered as strong, secure, and practically unforgeable forms of authentication. However, one aspect that is often overlooked is privacy. Initially a biometric can appear to be very private—your fingerprint, or iris, or voice, or gait belongs only to you. However, you do share it with the world, by handling objects, looking at the world, speaking, or walking, and once you share it with a database, that database now has a digital copy of it. Although the original biometric may be difficult to forge or steal, the digital image of it stored in a database may be vulnerable to viewing or reuse by unauthorized persons, or to theft for the purposes of a replay attack. And furthermore, if it is "stolen" or used in an unauthorized manner, it is almost impossible to change, unlike it would be if it were a simple password or a physical key or swipe card.

One solution to the privacy issue is the use of biometric encryption, also called biometric templates. This technique, proposed in 1994 by Tomko et al., stores only a template of the original biometric, instead of a verbatim digital copy of it. From a computer scientist's perspective, this is like storing the hash of a document, in a way similar to how passwords are stored in today's modern computer systems. This template storage approach has the additional advantage that, if the database is compromised, the existing template can be revoked and a new template created from the original biometric.

Previous work: Although there are many papers that explore gait as a conventional biometric, there are fewer that examine it as the input to a biometric encryption system. Previous work includes the work of Argyropoulos et al., who extract discriminative features from moving human silhouettes, and then use a biometric encryption scheme involving a channel encoding approach. Prior art also includes the work of Nickel et al., who use accelerometers to extract the gait data and use histogram generation to produce a biometric encryption scheme.

Summary of our approach: We propose what is, to the best of our knowledge, an entirely different approach, using a particular type of biometric encryption, a randomized biometric template (RBT), introduced by Ballard et al., and particularly suited for data, like gait data, that varies with time. We also select particular features of the gait data that we expect will provide as much entropy as possible so as to best distinguish samples and also best provide security.

Outline: Section 2 introduces biometric encryption, discusses the specific advantages and disadvantages of this approach. Section 3 provides our approach and discusses both the algorithm we are considering and the type of data we are using. Section 4 is the conclusion.

Biometric Encryption

As mentioned in Section 1, the concept of biometric encryption originated with Tomko et al., The idea behind it is that the biometric is bound with a digital key—either the biometric is used to generate the key (key generation), or the key is generated by other means and the biometric is combined with it (key binding). The resulting key is also called a template, and it is only this template—and not the original biometric—that is stored. This happens during the enrolment phase. When the user is successfully enrolled and presents him or herself at a later time for verification, in the key generation approach the biometric is used to generate the key again, or, in the key binding approach, the biometric is combined with the key again, and in either case the new result is compared with the result stored on enrolment. If they match, the credentials are accepted as valid. This is a process similar to one many password schemes use; however, the major difference is that the password the user types is supposed to be identical each time whereas the biometric data (possibly) changes slightly every time it is entered, so some sort of error correction is mandated. This kind of error correction is usually incorporated right into the algorithm.

Biometric encryption has been implemented with a number of different biometrics, including iris, fingerprints, voice, and handwriting. There have also been many algorithms introduced for biometric encryption, some of which have appeared in commercial products, e.g. PerSay for voice biometrics (now part of Nuance Communications). The first proposed algorithm was Mytec1 (Tomko et al., 1994); one of the most popular currently is fuzzy vault of Juels and Sudan (Juels and Sudan, 2002). Rathgeb and Uhl, and Cavoukian and Stoianov, each provide a survey of a number of biometric encryption algorithms and a discussion of the different biometrics that have been used with them.

Advantages and Disadvantages of Biometric Encryption

Biometric encryption has a number of advantages over conventional biometric systems and these advantages are enumerated in Cavoukian and Stoianov. The chief among these is the fact, mentioned above, that the actual biometric itself is never stored on the system. This means that if the system is hacked, all that is recovered is the output of the biometric encryption algorithm, and this output is not valid input to the algorithm so cannot be presented as a credential. Furthermore, a single individual biometric can be used to produce many biometric templates and this is an advantage if multiple systems are to be accessed, as a different credential can be created for each system (this is along the lines of the usual advice of "don't reuse your password on different systems."). Since the biometric encryption output is not identical to the biometric itself, it can be revoked or replaced if needed. This system can be integrated with other security systems in which the biometric encryption output can act as a key for a cryptosystem, or as a password for authentication, all without the need to memorize data.

There are performance differences between biometric encryption schemes and conventional biometric schemes. As Uludag et al., point out, in contrast to conventional biometric schemes, biometric encryption schemes require that the keys match perfectly to accept them as valid, whereas conventional biometric schemes allow for more leeway. Also, since in a conventional biometric the actual biometric can be manipulated and aligned to produce a higher acceptance rate, while the opportunity for such fine tuning is not available with biometric encryption schemes. For a fuller discussion of drawbacks see Uludag et al., Biometric encryption schemes, like any security scheme, are also vulnerable to a number of attacks, many of which involve brute force attacks, iteratively guessing the right answer, or exploiting nonuniformity in the data. Note, however, that conventional biometric schemes are vulnerable to most of the same attacks. See Cavoukian and Stoianov for a list of the major attacks.

A New Direction

Since there are a variety of different approaches to biometric encryption, we gave consideration to several algorithms before selecting one to focus on. We explored the feasibility of biometric encryption techniques fuzzy vault, (Juels and Sudan, 2002), fuzzy commitment (Juels and Wattenberg, 1999), and traditional biometric techniques cumulative match characteristic (CMC) (NIST, 2006), and hidden markov model (Kale, 2004), before settling on randomized biometric template (RBT) as an algorithm well-suited to the circumstances of gait.

Randomized Biometric Template

Ballard et al., introduce an approach called randomized biometric templates (RBTs) and describe both an algorithm and an approach to the input of the algorithm. One of the potential problems with biometric data is the lack of entropy in the data—without sufficient entropy there are not enough differences between users. There must be high inter-user variation for the biometric to be effective as a means of authentication. Ballard et al., propose taking account not just of the static biometric, but of how the features in the biometric are measured. This adds entropy to the measurement in process known as *hardening* the data, and it was proposed for biometric encryption by Monrose et al., In the case of Ballard et al., they select just some of the whole range of features to measure, and by varying this selection, they increase the entropy even more.

Which features they select also varies for each user, further increasing the randomness, since now the adversary must not only determine the biometric sample, but must also determine which features from that sample were used. They also select features that have high entropy in the population; that is, that vary a lot from user to user.

They also put an emphasis on *behavioural* biometrics rather than *physiological* ones. Behavioural biometrics relate to something a person does or an action a person performs, for example, keyboard strokes, voice patterns, gait. Because these can be changed more easily than physiological biometrics, they can be varied to get different keys, if required, or changed in the case of the loss or compromise of a key. Their algorithm has been implemented with handwriting data (Ballard et al., 2008) and voice data (Carrara and Adams, 2010), both of which are behavioural biometrics.

Their algorithm has two main parts. The first they call an enrolment algorithm. It is used to create the biometric template and a cryptographic key. The input to this algorithm is a sequence of biometric samples from the same user, along with an extra source of randomness (this source is optional). The second they call a key generation algorithm. It takes as input one biometric sample, the template, and the source of randomness. It is the job of the key generation algorithm to determine whether the one biometric sample and the template match, and, if so, outputs a cryptographic key which is the same as the cryptographic key output by the enrolment algorithm. Thus if the key from the enrolment algorithm was used to encrypt a plaintext, the key from the key generation algorithm could be used to decrypt it.

Carrara and Adams transfer the RBT algorithm of Ballard et al., from handwriting to voice data. In order to use the enrolment and key generation algorithms, however, they must first extract the appropriate features from the voice data through the use of a feature extraction algorithm that they designed. A key aspect of this feature extraction algorithm makes use of the fact that the data evolves through time, and, as gait has the same variability in time, it is a logical candidate for the same treatment. The challenge in applying it to gait will be to determine which features to use, and then to determine if these features provide sufficient entropy. Entropy measures the amount of uncertainty in the data. It can also be interpreted as measuring how different the elements of the data are from each other, or the variation in the data. The higher the entropy, the more random the data appears, and randomness in security data is obviously a desirable property. In terms of keys, if the set of keys displays high entropy then there are a lot of different keys for the adversary to choose from in trying to guess the real key. As the output of biometric encryption algorithms should be able to be used as the (128-bit) key in a current state-of-the-art cryptosystem, the amount of entropy displayed by a biometric encryption scheme should be sufficiently high (Rathgeb and Uhl, 2011). The entropy is to some degree a function of the features of the input. With respect to the RBT algorithm, Ballard et al., use the handwriting features such as start and end points of the signature and pen velocity, while Carrara and Adams must substantially process the voice data as reference points, or features, are not readily available (e.g. they split the data into frames, perform perceptual linear predictive analysis and delta analysis on each frame to produce 24 dimensional vectors consisting of these coefficients, etc.). In both cases they report the RBT algorithm generates keys that require the adversary to make at least guesses before guessing the right one for close to of users.

This measure is called the *guessing distance* and is simply, as it suggests, the number of guesses the adversary needs to recover a user's biometric features and/or the key. The assumption is that the attacker has full access to the user's biometric template (perhaps the adversary has hacked the system) but is trying to recover the features. This does require experiments, however, and statistical data since feature value distributions must be calculated (see Carrara and Adams for details). The alternative is *guessing entropy* used by Monrose et al., although, as pointed out by Carrara and Adams, it assumes a uniformity among users and that each key is equally likely to be guessable. Carrara and Adams suggest ways for further increasing entropy: samples from a broader range of users, and richer individual samples, i.e. samples with more variety in them and hence more features.

In terms of the security of biometric cryptosystems, one could also consider the false accept rate (FAR), which is the rate at which genuine keys are returned to bogus users, and false reject rate (FRR), which is the rate at which bogus keys are returned to genuine users (Rathgeb and Uhl, 2011). Of course these two quantities are intimately related, and decreasing one increases the other, and vice versa. The value of these rates as a measure of the efficacy of the biometric encryption system is clear; however, Buhan et al., have also shown there is a mathematical relationship between FAR, FRR, and min-entropy. Their work shows that in an ideal biometric encryption system, the min-entropy, m , must satisfy the relation FAR $\leq 2^{-m}$. The work of Vielhauer and Steinmetz further measures relationships between features (in handwriting) and thus, using these measures, one can identify which features are likely to produce desirable results, e.g. which features produce a high entropy. This is especially important in a system like RBT were one makes a choice of which features of the biometric to use in the algorithm. The aforementioned results and others are discussed in detail in Rathgeb and Uhl. Thus we are left with determining which features from a gait biometric to select, and with evaluating whether or not these features have sufficient entropy, FAR, and FRR to perform well with RBT. In order to answer these questions we need to consider the data.

The Data

Within the area of gait biometrics, there have been three main types of data capture systems used. These include Machine Vision based, Floor Sensor based and Wearable Movement Sensor based (see Gafurov, 2007 for a review). The Machine Vision and Wearable Movement Sensor based would be the two that would produce the data that we are proposing to use as a gait biometric. Video image techniques are used to extract features of recognition such as stride length, distance between head and pelvis, and distance between feet (Liu and Sarkar, 2004). Most past collection of gait biometric data has been based upon analysis of video-based figure silhouettes.

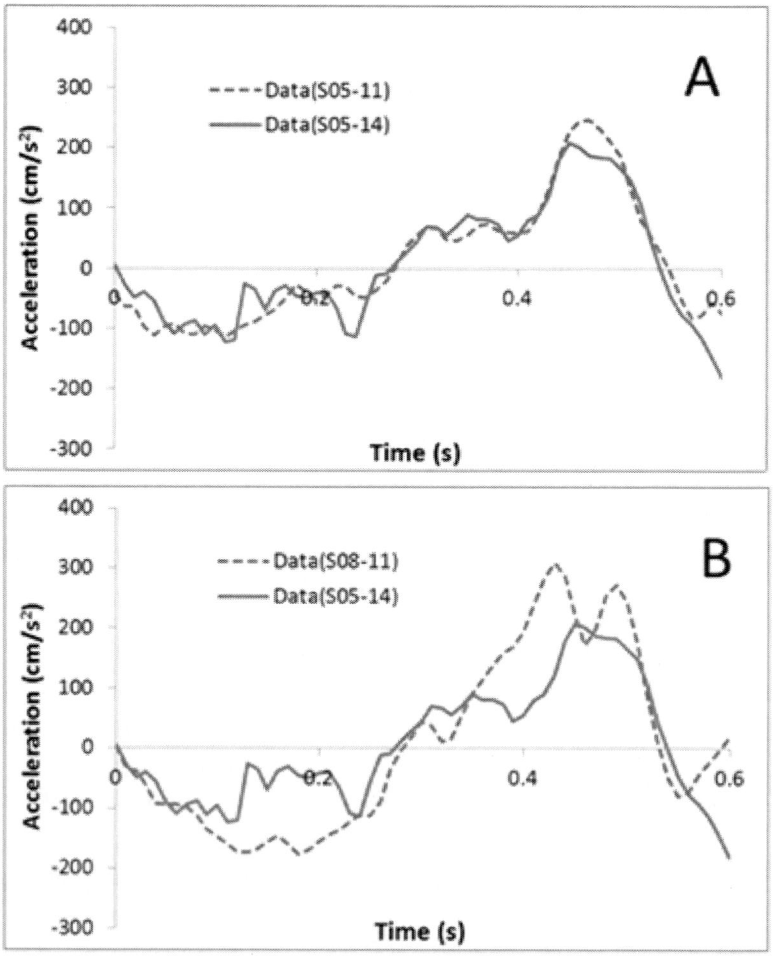

Figure 1. Acceleration profiles of the centre of mass during a gait cycle. A: Comparison between two trials of the same individual; B: Comparison between two individuals performing the same task.

The gait data that may present the most entropy for the biometric are parameters that describe the movement of the body's centre of mass. The body's centre mass can be easily tracked via video silhouette analysis during multiple gait cycles while also permitting the identification of the phases of gait in order to section the data into the specific phases of the gait cycle. One such measurement is the vertical acceleration of the centre of mass from the time of initial heel contact to the time of foot lift of that same foot. The analysis of this data is in its initial stages but Figure 1A illustrates the similarities between two separate trajectories of the same individual and then in Figure 1B the differences that exists between two different individuals performing the same task at approximately the same velocity.

The specific parameters have not be identified but rates of change, frequency content, maximums and minimums will all be considered for descriptive parameters. Additionally, most of the past research into the usage of gait movements as biometrics has classified mostly the motion in the forward-backward directions and we think the most unique patterns may actually occur in the off directions such as side-to-side or vertical. These off axis motions may be what constitutes the signature of an individual's gait and thus construct the entropy that is needed.

Conclusion

We have explored the technique of biometric encryption and the possibility of incorporating it into a biometric recognition scheme using gait data. Some systems using biometric encryption and gait data have been proposed; however, ours would use a biometric template particularly suited for dynamic data that varies with time as gait does, and would use novel features of the data to hopefully provide as much entropy as possible from the samples.

Acknowledgments

The authors were funded independently by NSERC Discovery Grants and by research grants from Wilfrid Laurier University. We thank D. Cox for preliminary investigations.

References

Argyropoulos, S., Tzovaras, D., Ioannidis, D., Strintzis, M. G. (2009). A channel coding approach for human authentication from gait sequences. *IEEE Transactions on Information Forensics and Security*, *4*(3):428-440.

Ballard, L., Kamara, S., Monrose, F., Reiter, M. K. (2008). Towards practical biometric key generation with randomized biometric templates, *Proceedings of the 15th ACM Conference on Computer and Communications Security, CCS*, 235–244.

Buhan, I. R., Doumen, J. M., Hartel, P. H., Veldhuis, R. N. J. (2006). Fuzzy extractors for continuous distributions. University of Twente, Technical Report

Carrara, B., Adams, C. (2010). You are the Key: generating cryptographic keys from voice biometrics, *Proceedings of the 8th Annual Conference on Privacy, Security and Trust (PST 2010)*, 213–222.

Cavoukian, A., Stoianov, A. (2009). Biometric encryption, in *Springer Encyclopedia of Biometrics*, S. Z. Li, ed.

Gafurov, D. (2007). A survey of biometric gait recognition: approaches, security, and challenges, *Proceedings of Annual Norwegian Computer Science Conference*, Tapir, 119-130.

Juels, A., Sudan, M. (2002). A fuzzy vault scheme. *Proc 2002 IEEE Int Symp on Information Theory*, 408.

Juels, A., Wattenberg M. (1999). A fuzzy commitment scheme, in G. Tsudik, ed., *6th ACM Conference on Computer and Communications Security*, 28-36.

Kale, A. E. (2004). Identification of humans using gait. *IEEE Transaction on Image Processing,* 1163–1172.

Liu, Z. and S. Sarkar S. (2004). Simplest representation yet for gait recognition: averaged silhouette, *International Conference on Pattern Recognition*, 211-214.

Monrose, F., Reiter, M. K., and Wetzel, S. (2002). Password Hardening based on Keystroke Dynamics. *International Journal of Information Security*, *1*(2):69–83.

Nickel, C., Zhou, X., Busch, C. (2010). Template protection for biometric gait data, in Bršmme, Arslan (Ed.); Busch, Christoph (Ed.), BIOSIG 2010, Special Interest Group on Biometrics and Electronic Signatures. Proceedings, *Lecture Notes in Informatics (LNI) - Proceedings 164*, 73-81

NIST - Subcommittee on Bioinformatics. (2006, August 7). Biometrics Testing and Statistics. Retrieved December 20, 2012, from http://www.biometrics.gov/Documents/BioTestingAndStats.pdf

Rathgeb, C., Uhl, A. (2011). A survey on biometric cryptosystems and cancelable biometrics, *EURASIP Journal on Information Security, 3*.

Tomko, G. J., Soutar, C., Schmidt, G. J. (1996). Fingerprint controlled public key cryptographic system. U.S. Patent 5541994, (Filing date: Sept. 7, 1994).

Uludag, U., Pankanti, S., Prabhakar, S., Jain, A. K. (2004). Biometric cryptosystems: issues and challenges, *Proc. IEEE 92*(6):948–960.

Vielhauer, C., Steinmetzm, R. (2004). Handwriting: feature correlation analysis for biometric hashes, *EURASIP J. Appl. Signal Process,* (1):542-558.

Editors' Contact Information

Dr. Li Li
Research Professor
Department of Health and Kinesiology
Georgia Southern University
PO Box 8076
Statesboro, GA 30460
Tel: 912-478-8015
lili@georgiasouthern.edu

Mr. Matthew Holmes
Department of Health and Kinesiology
Georgia Southern University
PO Box 8076
Statesboro, GA 30460
Tel: 912-478-8015

Index

A

B

C

D

G

H

I

N

O

P

T

U

V

W

Y